AF352830

Storm Tide and Wave Simulations and Assessment II

Storm Tide and Wave Simulations and Assessment II

Editors

Shih-Chun Hsiao
Wen-Son Chiang
Wei-Bo Chen

MDPI • Basel • Beijing • Wuhan • Barcelona • Belgrade • Manchester • Tokyo • Cluj • Tianjin

Editors

Shih-Chun Hsiao
Department of Hydraulic and
Ocean Engineering, National
Cheng Kung University
Taiwan

Wen-Son Chiang
Tainan Hydraulics Laboratory,
National Cheng
Kung University
Taiwan

Wei-Bo Chen
National Science and Technology
Center for Disaster Reduction
Taiwan

Editorial Office
MDPI
St. Alban-Anlage 66
4052 Basel, Switzerland

This is a reprint of articles from the Special Issue published online in the open access journal *Journal of Marine Science and Engineering* (ISSN 2077-1312) (available at: https://www.mdpi.com/journal/jmse/special_issues/Wei-Bo_storm_tide_wave_simulations_assessment).

For citation purposes, cite each article independently as indicated on the article page online and as indicated below:

LastName, A.A.; LastName, B.B.; LastName, C.C. Article Title. *Journal Name* **Year**, *Volume Number*, Page Range.

ISBN 978-3-0365-3567-8 (Hbk)
ISBN 978-3-0365-3568-5 (PDF)

Contents

Journal of
*Marine Science
and Engineering*

Editorial

Storm Tide and Wave Simulations and Assessment II

Shih-Chun Hsiao [1], Wen-Son Chiang [2] and Wei-Bo Chen [3,*]

[1] Department of Hydraulic and Ocean Engineering, National Cheng Kung University, Tainan 701, Taiwan; schsiao@mail.ncku.edu.tw
[2] Tainan Hydraulics Laboratory, National Cheng Kung University, Tainan 70101, Taiwan; chws@mail.ncku.edu.tw
[3] National Science and Technology Center for Disaster Reduction, New Taipei 231, Taiwan
* Correspondence: wbchen@ncdr.nat.gov.tw; Tel.: +886-2-8195-8611

1. Introduction

The storm tides, surges, and waves that are associated with typhoons/tropical cyclones/hurricanes are the most severe threats to coastal zones, nearshore waters, and navigational safety. Therefore, predicting typhoon/tropical cyclone/hurricane-induced storm tides, surges, waves, and coastal erosion is essential for reducing the loss of human life and property as well as for mitigating coastal disasters. There is still a growing demand for novel techniques that could be adopted to resolve the complex physical processes of storm tides, surges, waves, and coastal erosion, even if many studies on the hindcasting/predicting/forecasting of typhoon-driven storm tides, surges, and waves as well as on morphology evolution have been carried out through the use of numerical models in the last decade.

The intention of this Special Issue was to collect the latest studies on storm tide, surge, and wave modeling and analysis utilizing dynamic and statistical models and artificial intelligence approaches to improve our simulation and analytic capabilities and our understanding of storm tides, surges, and waves. Five high-quality papers were accepted for publication in this Special Issue; these papers cover the application and development of many high-end techniques that can be used to study storm tides, surges, waves as well as for on-site investigations of coastal erosion and accretion.

2. Details of Papers

To analyze wave-induced effects on sea surface temperature simulations during typhoon events, Sun et al. [1] investigated sea surface temperature (SST) cooling under binary typhoon conditions. They mainly focused on parallel- and cross-type typhoon paths during four typhoon events: Tembin and Bolaven in 2012 and Typhoon Chan-hom and Linfa in 2015. Wave-induced effects were simulated using a third-generation numeric model, WAVEWATCH III (WW3), and were subsequently included in SST simulations using the Stony Brook Parallel Ocean Model (sbPOM). They indicated that the vertical profile of the SST simulation indicated that the disturbance depth increased (up to 100 m) for cross-type typhoon paths because the mixing intensity was greater for the cross-type typhoons than it was for the parallel-type typhoons.

To assess the effects of depth-induced breaking on wind wave simulations in the shallow nearshore waters off of northern Taiwan during the passage of two super typhoons using a high-resolution coupled wave-circulation model, Hsiao et al. [2] found that the significant wave height that was induced by the typhoon winds in the surf zone were more sensitive to the different wave-breaking formulations that were used in the wave–circulation model. The significant wave height simulations in the surf zone were insensitive to the wave-breaking criterion (γ) during the passage of typhoons. In shallow nearshore waters, the utilization of a constant γ for the wave–circulation model always

Citation: Hsiao, S.-C.; Chiang, W.-S.; Chen, W.-B. Storm Tide and Wave Simulations and Assessment II. *J. Mar. Sci. Eng.* **2022**, *10*, 379. https://doi.org/10.3390/jmse10030379

Received: 24 February 2022
Accepted: 1 March 2022
Published: 6 March 2022

Publisher's Note: MDPI stays neutral with regard to jurisdictional claims in published maps and institutional affiliations.

produced peak significant wave heights that are smaller than those that were achieved when γ based on local steepness or peak steepness was used.

To evaluate the focusing waves induced by Bragg resonance with a V-shaped undulating bottom, Zhang et al. [3] proposed sinusoidal sandbars with a horizontal V-shaped pattern that was formed by two continuous undulating bottoms that were inclined towards each other at an angle with the center axis perpendicular to the shoreline. Based on the high-order spectral (HOS) numerical model, both the Bragg resonance characteristics that were induced by the regular waves and random waves were investigated. With regular waves, the study showed that the wave-focusing effect is related to the angle of the V-shaped undulating bottom and that the optimal angle of inclination for the V-shaped undulating bottom is 162.24°

To derive the 100-year total water level around the coast of Corsica by combining trivariate extreme value analysis and coastal hydrodynamic models, Louisor et al. [4] focused on providing extreme scenarios through which wind wave and coastal hydrodynamic models, i.e., SWAN and SWASH-2DH, could be populated in order to compute the 100-year total water level (100y-TWL) along the coasts. They showed how the proposed multivariate extreme value analysis can help to more accurately define low-lying zones that are potentially exposed to coastal flooding, especially in Corsica, where a unique value of 2 m was taken into account in previous studies. The computed 100y-TWL values were determined to be between 1 m along the eastern coasts and 1.8 m on the western coast. The calculated values are also below the 2.4 m threshold that is recommended when considering sea level rise (SLR). Their results highlight the added value of performing a complete integration of extreme offshore conditions together with their dependence on hydrodynamic simulations for screening out the coastal areas that may potentially be exposed to flooding.

For the on-site investigation of coastal erosion and accretion for the northeast region of Taiwan, Liang et al. [5] investigated coastal erosion and accretion in that area through a series of on-site surveys. The results of the bathymetric surveys suggest that the shoreline of Yilan County tends to accrete during the summer because of abundant sediment from the rivers; however, the shoreline is eroded in winter due to the large waves that are induced by the northeast monsoon. Additionally, the calculated net volume of erosion and accretion between each pair of cross sections showed that the length of the coastline that impacted by estuarine sediment transport is approximately 2 km long and spans from north to south along the Lanyang River estuary coastline.

Funding: This research received no external funding.

Conflicts of Interest: The authors declare no conflict of interest.

References

1. Sun, Z.; Shao, W.; Yu, W.; Li, J. A Study of Wave-Induced Effects on Sea Surface Temperature Simulations during Typhoon Events. *J. Mar. Sci. Eng.* **2021**, *9*, 622. [CrossRef]
2. Hsiao, S.-C.; Wu, H.-L.; Chen, W.-B.; Guo, W.-D.; Chang, C.-H.; Su, W.-R. Effect of Depth-Induced Breaking on Wind Wave Simulations in Shallow Nearshore Waters off Northern Taiwan during the Passage of Two Super Typhoons. *J. Mar. Sci. Eng.* **2021**, *9*, 706. [CrossRef]
3. Zhang, H.; Tao, A.; Tu, J.; Su, J.; Xie, S. The Focusing Waves Induced by Bragg Resonance with V-Shaped Undulating Bottom. *J. Mar. Sci. Eng.* **2021**, *9*, 708. [CrossRef]
4. Louisor, J.; Rohmer, J.; Bulteau, T.; Boulahya, F.; Pedreros, R.; Maspataud, A.; Mugica, J. Deriving the 100-Year Total Water Level around the Coast of Corsica by Combining Trivariate Extreme Value Analysis and Coastal Hydrodynamic Models. *J. Mar. Sci. Eng.* **2021**, *9*, 1347. [CrossRef]
5. Liang, T.-Y.; Chang, C.-H.; Hsiao, S.-C.; Huang, W.-P.; Chang, T.-Y.; Guo, W.-D.; Liu, C.-H.; Ho, J.-Y.; Chen, W.-B. On-Site Investigations of Coastal Erosion and Accretion for the Northeast of Taiwan. *J. Mar. Sci. Eng.* **2022**, *10*, 282. [CrossRef]

Journal of
*Marine Science
and Engineering*

Article

A Study of Wave-Induced Effects on Sea Surface Temperature Simulations during Typhoon Events

Zhanfeng Sun [1], Weizeng Shao [1,2,*], Wupeng Yu [3] and Jun Li [4]

[1] College of Marine Sciences, Shanghai Ocean University, Shanghai 201306, China; zfsun@shou.edu.cn
[2] National Satellite Ocean Application Service, Ministry of Natural Resources, Beijing 100081, China
[3] Marine Science and Technology College, Zhejiang Ocean University, Zhoushan 316022, China; y_wupeng@outlook.com
[4] East China Sea Environment Monitoring, Ministry of Natural Resources, Shanghai 201306, China; Lij@ecs.mnr.gov.cn
* Correspondence: shaoweizeng@mail.tsinghua.edu.cn; Tel.: +86-21-6190-0326

Citation: Sun, Z.; Shao, W.; Yu, W.;
Li, J. A Study of Wave-Induced
Effects on Sea Surface Temperature
Simulations during Typhoon Events.
J. Mar. Sci. Eng. **2021**, *9*, 622. https://
doi.org/10.3390/jmse9060622

Academic Editors: Wei-Bo Chen,
Shih-Chun Hsiao
and Wen-Son Chiang

Received: 23 April 2021
Accepted: 2 June 2021
Published: 3 June 2021

Abstract: In this work, we investigate sea surface temperature (SST) cooling under binary typhoon conditions. We particularly focus on parallel- and cross-type typhoon paths during four typhoon events: Tembin and Bolaven in 2012, and Typhoon Chan-hom and Linfa in 2015. Wave-induced effects were simulated using a third-generation numeric model, WAVEWATCH III (WW3), and were subsequently included in SST simulations using the Stony Brook Parallel Ocean Model (sbPOM). Four wave-induced effects were analyzed: breaking waves, nonbreaking waves, radiation stress, and Stokes drift. Comparison of WW3-simulated significant wave height (SWH) data with measurements from the Jason-2 altimeter showed that the root mean square error (RMSE) was less than 0.6 m with a correlation (COR) of 0.9. When the four typhoon-wave-induced effects were included in sbPOM simulations, the simulated SSTs had an RMSE of 1 °C with a COR of 0.99 as compared to the Argos data. This was better than the RMSE and COR recovered between the measured and simulated SSTs, which were 1.4 °C and 0.96, respectively, when the four terms were not included. In particular, our results show that the effects of Stokes drift, as well as of nonbreaking waves, were an important factor in SST reduction during binary typhoons. The horizontal profile of the sbPOM-simulated SST for parallel-type typhoon paths (Typhoons Tembin and Bolaven) suggested that the observed finger pattern of SST cooling (up to 2 °C) was probably caused by drag from typhoon Tembin. SST was reduced by up to 4 °C for cross-type typhoon paths (Typhoons Chan-hom and Linfa). In general, mixing significantly increased when the four wave-induced effects were included. The vertical profile of SST indicated that disturbance depth increased (up to 100 m) for cross-type typhoon paths because the mixing intensity was greater for cross-type typhoons than for parallel-type typhoons.

Keywords: typhoon wave; sea surface temperature; WAVEWATCH-III; sbPOM

1. Introduction

Typhoons occur frequently in the Western Pacific Ocean (WP) [1], affecting energy exchange at the air–sea boundary layer [2,3] and leading to several secondary hazards, such as extreme waves [4,5], landslides, and heavy rains [6]. Binary typhoons, in which two storms of tropical cyclone intensity or more occur simultaneously, have also been recorded in the WP. Because of the strong wind interactions endemic to binary typhoons, binary typhoons have more complicated effects on the sea surface than single typhoons due to the influence of total heat flux exchange on the upper ocean response [7,8]. At present, moored buoys [9] and satellites [10,11] provide real-time observations of oceanic conditions, particularly winds and waves, during hurricanes and typhoons. However, those devices are unable to generate time-series data with a fine spatial resolution; that is, the resolution of a scatterometer is typically 12.5 km [12], while that of an altimeter is 10 km [13]. Thus, the data collected by these devices are not sufficient for long-term distribution analyses.

Over recent decades, as wave theory and computation technologies have matured, several numeric wave models have been proposed. At the beginning of the 1980s, WAMDIG initially proposed the third-generation numeric wave ocean model (WAM) [14], which integrated basic wave propagation effects to describe the evolution of a two-dimensional ocean wave spectrum. Subsequent authors developed the WAVEWATCH-III (WW3) [15] and Simulating Waves Nearshore (SWAN) [16] models based on the principles of the WAM. The main difference between the WW3 and SWAN models is the applicability of the model: SWAN was originally developed as a nearshore model, while WW3 was developed for the oceanic scales. Therefore, the WW3 model is usually employed for wave simulations over large regions, such as global seas [17] or the western Pacific Ocean [18], while the SWAN model is typical used to analyze coastal waters [19]. Model-simulated waves are also commonly used in synthetic aperture radar (SAR) wave monitoring [20,21] and, in particular, as auxiliary data for typhoon analysis [22].

During typhoon events, the sea state is complicated due to strong, synoptic-scale air–sea interactions and turbulent mixing at the sea surface [23]. In particular, sea surface temperatures (SSTs) during typhoons are rarely measured using real-time techniques such as Argos [24]. This air–sea mixing, as well as the Ekman pumping induced by cyclonic vorticity, deepens the mixed layer at the sea surface, decreasing SST [25–27]. Remarkably, observational data have shown a maximum SST cooling of 9 °C [28], and short-term satellite data recorded during two typhoon events revealed anomalously cold SST patches that were up to 6 °C colder than the SSTs of the surrounding warm tropical sea [29]. SST cooling in turn increases typhoon intensity and movement by modulating energy fluxes and stability at the air–sea boundary layer [30]. Therefore, SST cooling is one of the more noticeable oceanic responses to a moving typhoon due to its significant influence on oceanic and atmospheric dynamics [31]. Due to the limited availability of observational data during typhoons, the coupled atmospheric–oceanic model provides a powerful alternative way to study SST cooling and its impact on typhoon intensity [32].

Recent numerical experiments have aimed to clarify the unique characteristics and underlying mechanisms of SST cooling [33,34]. Two case studies [35,36] have suggested that SST cooling in the inner-core region of the cyclone, which is defined as within a 111-km radius of the cyclone center [37], may weaken typhoon intensity. However, the largest SST reduction often occurs in the right-rear quadrant of the typhoon. Although sea-surface waves themselves act over small scales, ranging from meters to kilometers, wave-induced effects, such as breaking waves, nonbreaking waves, radiation stress, and Stokes drift, affect the air–sea energy exchange at the boundary layer, especially during strong winds. Thus, wave-induced effects should be considered in analyses of SST cooling. The produced cooling is a function of both typhoon forward speed and intensity. Generally, the lower the forward speed, the higher the cooling rate and the higher the intensity, meaning a larger cooling rate is expected [38,39]. The extra cooling and turbulent mixing on the right side of the track in the Northern Hemisphere as a result of the rightward bias can contribute to a larger deepening of the mixed layer [40]. In most cases, binary typhoons are stronger than single typhoons in terms of duration and range, and will cause strong upwelling and mesoscale cyclone vortices in certain areas, which will also have a greater impact on SST [41]. Furthermore, it is important to assess SST cooling during binary typhoons, which include both parallel- and cross-type movements.

The ocean circulation model, which has been termed the Princeton Ocean Model (POM), is commonly used to simulate global marine dynamics [42,43], such as current and SST. An updated version of POM, the Stony Brook Parallel Ocean Model (sbPOM) [44], has been improved and enhanced using the parallel computation technique. The scalability of the POM model is better than that of its predecessors. In this study, we simulated wave fields during certain typhoon events using the WW3 model and calculated four of the effects of strong winds: breaking waves, nonbreaking waves, radiation stress, and

Stokes drift. Subsequently, SST was simulated using an sbPOM that included these four factors. In particular, we focused on fluctuations in SST cooling during various types of typhoon movement.

Indeed, the primary aim of this study was to assess SST cooling during binary typhoons with different paths. The datasets used, which are described in Section 2, include the typhoon events, the forcing wind fields, the open boundary conditions for modeling, and measurements from the Jason-2 altimeter and Argos. The model settings for the WW3 and sbPOM simulations are also given in Section 2. Waves were simulated using the WW3 model and SSTs were simulated using sbPOM, both based on the four effects of the typhoon waves. We assessed the accuracies of these models and the discussions in Section 3. Our conclusions are summarized in Section 4.

2. Materials and Methods

Four typhoons that passed through the China Sea were analyzed in this study: Tembin, Bolaven, Chan-hom, and Linfa. The tracks of these typhoons were identified as either parallel- or cross-type. Superimposition of the tracks of these four typhoons over the water depths obtained from the General Bathymetry Chart of the Oceans (GEBCO) data showed that the paths of Tembin and Bolaven (19–30 August 2012) were almost parallel, while the path of Chan-hom intersected that of Linfa (7–12 July 2015; Figure 1).

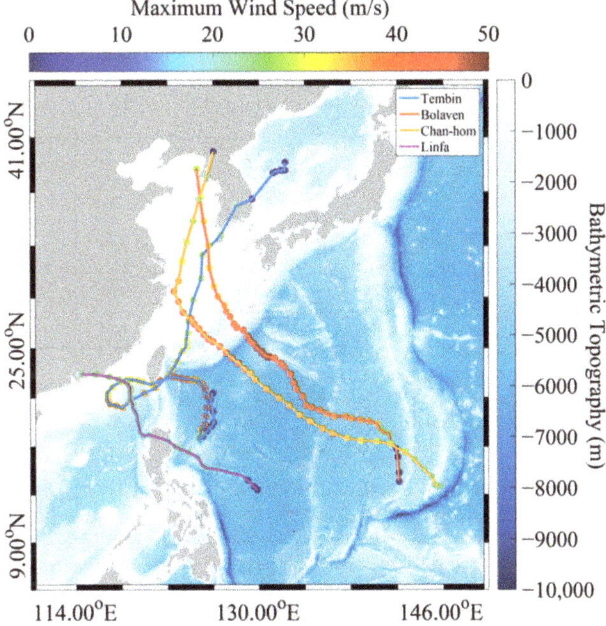

Figure 1. The tracks of the typhoons analyzed in this study superimposed on the water depth data from the General Bathymetry Chart of the Oceans (GEBCO).

Two numeric models, a WW3 model and a sbPOM, were implemented for the China Seas. The European Centre for Medium-Range Weather Forecast (ECMWF) has provided operational products for the global seas since 1958; ECMWF data have a high resolution (up to 0.125°) at intervals of 1 h. Using the ECMWF wind data, we hindcasted the long-term wave distributions previously simulated using SWAN and WW3 models in two Chinese seas: the Bobai Sea [45] and the South China Sea [46]. The results indicated that the model-simulated waves were consistent with buoy data and altimeter measurements. However, simulated measurements were systematically underestimated as compared to observational measurements [47]. In a recent study, we reconstructed "H-E winds," composited of ECMWF winds and a parametric Holland model [15], for typhoons. The model was trained by fitting the shape parameter to buoy-measured observations. We then compared simulated wind speeds with those measured by moored buoys, and found that the root mean square error (RMSE) of wind speed was less than 3 m/s for the shape parameter equivalent to 0.4 [15]. A representative example of H-E wind fields, for Typhoons Chan-hom and Linfa at 18:00 UTC on 5 July 2015, is shown in Figure 2. In this figure, two typhoon centers are clearly apparent in the H-E wind fields. Importantly, the underestimation of ECMWF winds was improved by reanalysis using H-E winds.

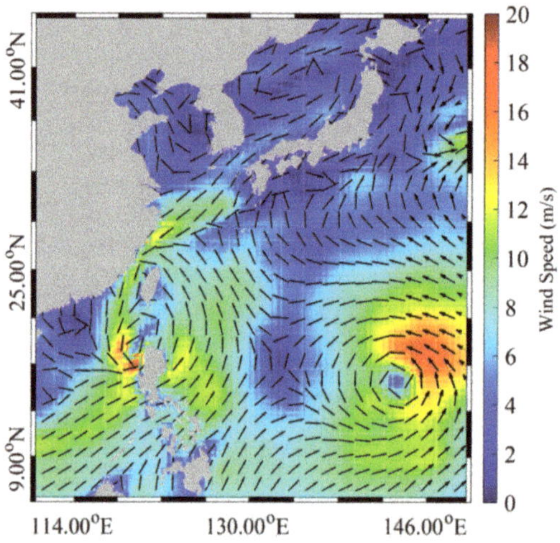

Figure 2. The wind map for Typhoons Chan-hom and Linfa at 18:00 UTC on 5 July 2015. Wind maps were composited using the European Centre for Medium-Range Weather Forecasts (ECMWF) wind data and a parametric Holland model (H-E).

The initial fields for the SST simulation of sbPOM were the monthly average SST and the sea surface salinity from the Simple Ocean Data Assimilation (SODA) data, which have a spatial grid resolution of 0.5°. A representative example from August 2012 is shown in Figure 3. The upper boundary forcing fields were obtained based on the National Centers for Environmental Prediction (NCEP) total heat flux parameters (latent heat flux, sensible heat flux, long-wave radiation, and short-wave radiation) at 6 h intervals and a spatial resolution of 1.875° × 1.905° (longitude × latitude); in contrast, ECMWP provides flux data twice per day. Argos is an international cooperative project begun in 2000. The Argos project, which aims to profile ocean temperature and salinity, is a major component of many ocean observation systems [48]. In this study, we used the high-quality SST measurements from Argo project to validate the simulations of the sbPOM. As an example, the map of NCEP total heat flux at 18:00 UTC on 5 July 2015 is shown in Figure 4; in this figure, triangles represent the geographic locations of the available Argos stations (>20) used in this study.

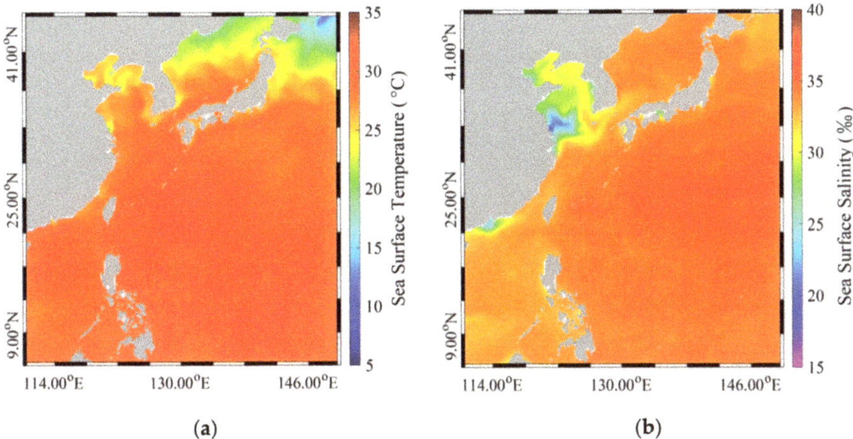

Figure 3. The Simple Ocean Data Assimilation (SODA) data used in the Stony Brook Parallel Ocean Model (sbPOM) on August 2012. (a) Monthly average sea surface temperature. (b) Monthly average sea surface salinity.

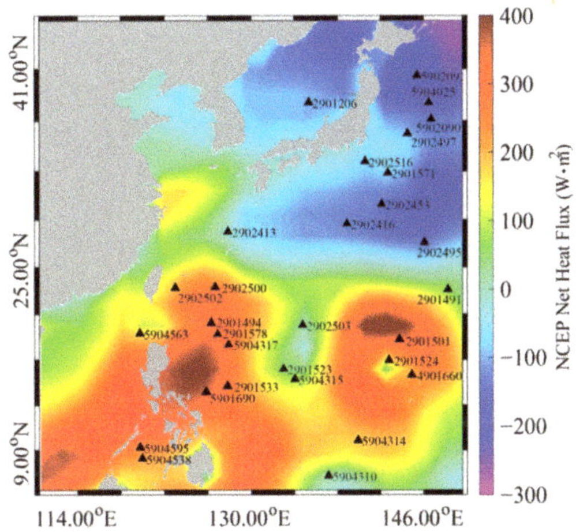

Figure 4. The National Centers for Environmental Prediction (NCEP) total heat flux at 18:00 UTC on 5 July 2015, overlaid with the geographic locations of the Argos stations used in this study.

In addition, SWH measurements were obtained from the altimeter Jason-2, which is a follow-on satellite to that of the Jason-1 oceanography mission of the National Aeronautics and Space Administration (NASA) and is a valuable source of global wave distributions [49]. These SWH data were used to validate the WW3-simulated SWH data. The WW3-simulated SWH map was overlaid the data from the Jason-2 altimeter at 18:00 UTC on 5 July 2015 (Figure 5). In this figure, the two cyclone-induced wave patterns are clearly visible. The basic settings of the WW3 model and the sbPOM (e.g., forcing fields, open boundary conditions, and output resolution) are summarized in Table 1. The details of the WW3 model and the sbPOM are given in Appendices A and B, respectively.

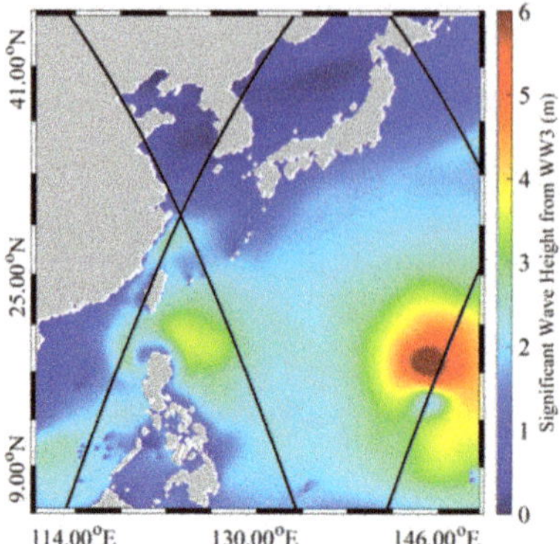

Figure 5. The significant wave height (SWH) data obtained from the WAVEWATCH-III (WW3) model at 18:00 UTC on 5 July 2015, overlaid with the footprints of the Jason-2 altimeter.

Table 1. The basic settings for the WAVEWATCH-III (WW3) model and the Stony Brook Parallel Ocean Model (sbPOM).

	Forcing Fields	Output Resolution	Open Boundary Conditions
WW3	Winds composed using the European Centre for Medium-Range Weather Forecast (ECMWF) data and a parametric Holland model (H-E)	Temporal resolution of 30 min and spatial grid resolution of 0.1°	/
sbPOM	H-E winds; Simple Ocean Data Assimilation (SODA) sea surface temperature and salinity Wave-induced: breaking wave; nonbreaking wave Radiation stress; stokes drift	Temporal resolution of 30 min and spatial grid resolution of 0.25°	National Centers for Environmental Prediction (NCEP) latent heat flux NCEP sensible heat flux NCEP long-wave radiation NCEP short-wave radiation

3. Results and Discussions

3.1. Validation of the SWHs Simulated Using the WW3 Model

The wave fields during the four typhoon events were simulated using WW3 model. The WW3-simulated SWHs were compared with the SWH measurements of the Jason-2 altimeter statistically using root mean squared error (RMSE) and correlations (CORs). RMSE and COR were calculated as follows:

$$RMSE = \sqrt{\frac{\sum_{i-1}^{N}\left(X_{mod}^{i} - X_{obs}^{i}\right)^2}{N}} \text{ and} \tag{1}$$

$$COR = \frac{Cov(X_{mod}, X_{obs})}{\sqrt{Var(X_{mod})Var(X_{obs})}}, \tag{2}$$

where N was the number of matchups, X_{mod} were the model-simulated results, X_{obs} were the measurements from the Jason-2 altimeter, Cov was the covariance, and Var was the variance.

The analysis of more than 10,000 matchups during Typhoons Tembin and Bolaven (19–30 August 2012) resulted in SWHs with an RMSE of 0.54 and a COR of 0.91 (Figure 6a). Similarly, the SWHs of Typhoons Chan-hom and Linfa (2–11 July 2015) had an RMSE of 0.50 m and a COR of 0.90 (Figure 6b). These results indicate that the WW3-simulated wave fields during typhoons were reliable, even under extreme conditions (SWH > 8). The methods used to calculate breaking waves, nonbreaking waves, radiation stress, and Stokes drift are given in Appendix A using WW3-simulated parameters (e.g., SWH, mean wave period, wavelength, and dominant wave propagation velocity).

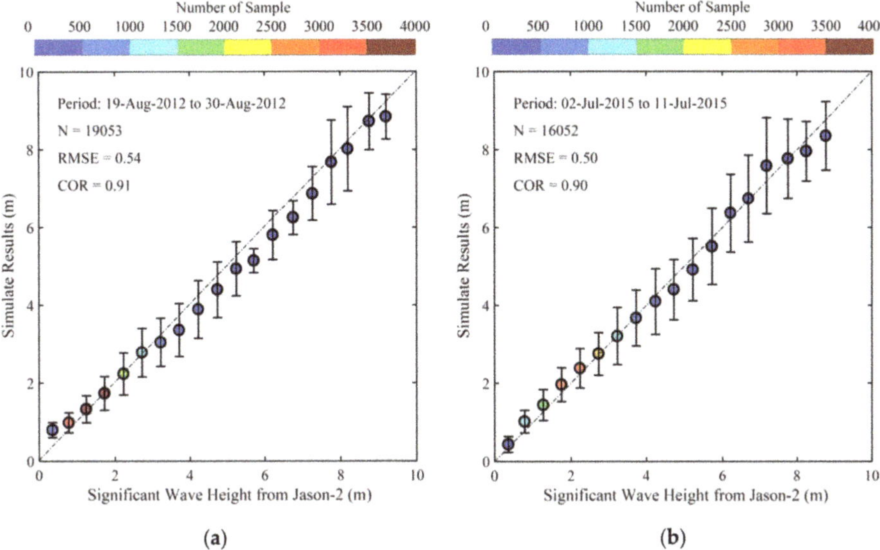

Figure 6. Comparisons of WW3-simulated SWHs with measurements from the Jason-2 altimeter in 0.5-m bins between 0 and 10 m. (**a**) 19–30 August 2012; (**b**) 2–11 July 2015.

3.2. Analysis of SSTs Simulated Using the sbPOM

The four effects, which were empirically calculated based on wave parameters derived from the WW3 model, were treated as forcing fields in the sbPOM. Here, we first present the results of the sbPOM, including the individual wave-induced terms simulated using the WW3 model. When the simulated results were compared with the Argos measurements during the four typhoons, the bias in nonbreaking stress and Strokes drift terms was about 1 °C; the bias in the breaking and radiation stress terms was greater (Table 2).

Table 2. Statistical comparisons of sea surface temperatures (SSTs) generated by the sbPOM and Argos during the four typhoons.

	Wave Breaking	**Nonbreaking Wave**	**Radiation Stress**	**Stokes Drift**
RMSE (°C)	1.23	1.16	1.40	1.02
COR	0.96	0.97	0.97	0.98

We included Stokes drift, accounting for depth decay, in the sbPOM. The interaction between friction velocity and Stokes drift may result in Langmuir circulations in binary typhoons, and may subsequently affect SST. This is probably why Stokes drift was the least biased term.

We comparted the SSTs from the available Argos data with those simulated using the sbPOM in 2 °C bins between 0 and 30 °C during the Tembin and Bolaven typhoons (19–30 August 2012; Figure 7). Figure 7a shows SSTs simulated without the four effects induced by typhoon waves (RMSE of 1.42 °C and COR of 0.95), while Figure 7b shows SSTs simulated including the four effects induced by typhoon waves (RMSE of 1.07 °C and COR of 0.99). Statistical analyses of the Chan-hom and Linfa typhoons (2–11 July 2015) yielded similar results: the results of the analysis including the four effects induced by typhoon waves (RMSE of 1.01 °C and COR of 0.99; Figure 8a) were better than those not including the four effects induced by typhoon waves (RSME of 1.37 °C and COR of 0.96; Figure 8b). Therefore, we concluded that the accuracy of the sbPOM SST simulation was improved when the four WW3-simulated effects induced by typhoon waves were included. The RMSE of SST was 1.40 °C when radiation stress was included, which was similar to the results generated without including any wave-induced effects. This suggested that radiation stress had little influence on SST cooling. Nonbreaking wave-induced mixing alone led to SST cooling in a study of individual typhoons [50]. However, our results show that effects of Stokes drift were also an important factor in SST cooling during binary typhoons.

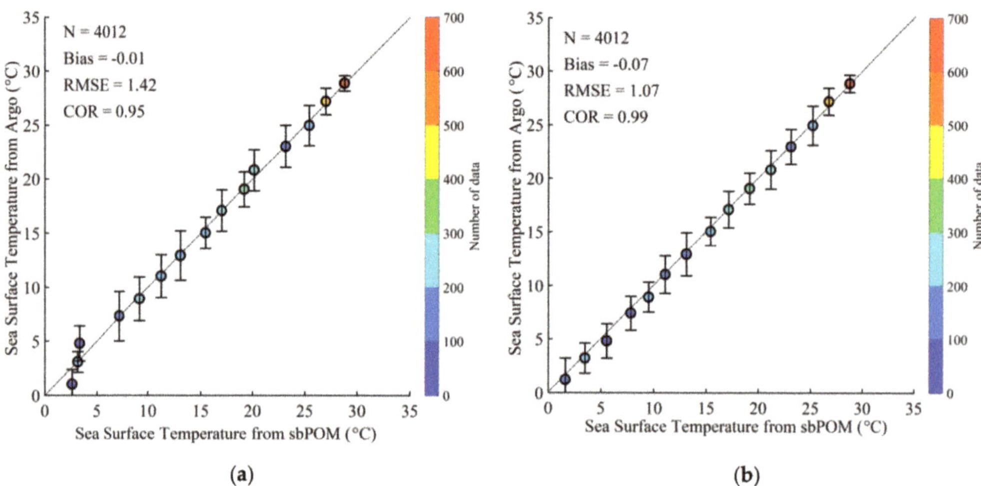

Figure 7. Comparisons of the SSTs from Argos with SSTs simulated using the sbPOM for 2 °C bins between 0 and 30 °C during the period 19–30 August 2012. (**a**) Not including the four effects induced by typhoon waves. (**b**) Including the four effects induced by typhoon waves.

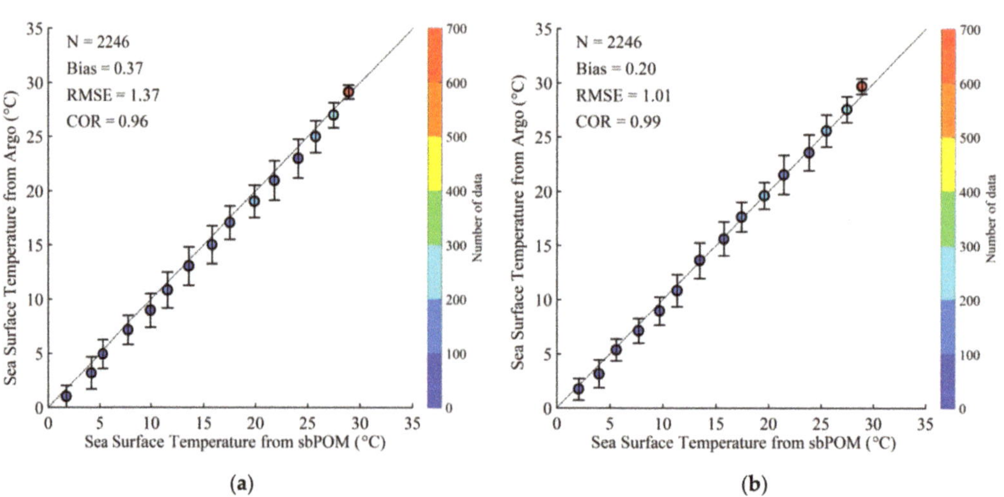

Figure 8. Comparisons of the SSTs from Argos with SSTs simulated using the sbPOM for 2 °C bins between 0 and 30 °C during the period 2–11 July 2015. (**a**) Not including the four effects induced by typhoon waves. (**b**) Including the four effects induced by typhoon waves.

3.3. Discussions

Several cases of typhoon-associated SST cooling have been analyzed using numeric modeling [51] and satellite observations [52]. These previous studies have shown that the paths of binary typhoons can be divided into six categories, which are also related to the intensity of the airflow [53]. SST cooling occurs to the right of the typhoon path in the Northern Hemisphere, and the return of SST to normal depends on the thickness of the upper ocean layer and the wind conditions [54]. These findings suggest that SST cooling patterns will be more complicated during binary typhoons, especially simultaneous typhoon events, due to the interactions between the two wind and wave systems.

The daily average SST distributions during the Tembin and Bolaven typhoons (red lines) on 26–30 August 2012 are shown in Figure 9a–d. Similarly, Figure 10 shows the daily average SST distributions during the Chan-hom and Linfa typhoons on 7–11 July 2015. The reduction in SST associated with cross-type typhoon paths (Figure 9b,c) was up to 4 °C, while the reduction in SST associated with parallel-type typhoon paths was up to 2 °C. The finger pattern of SST cooling shown in Figure 9a was probably caused by drag from Typhoon Tembin. The mixing associated with the selected typhoons, including four effects, is shown at Site A (Figure 11a) and at Site B (Figure 11b). In these figures, K_h represents the mixing induced by heat flux and K_m represents the mixing induced by momentum. Generally, mixing significantly increased during parallel-type typhoons when the four wave-induced effects were included. In particular, mixing intensity up to a depth of 50 m was greater for cross-type typhoons (maximum 0.2) than for parallel-type typhoons (maximum 0.1). This suggested that the strong energy exchange associated with cross-type typhoon paths led to the substantial reduction in SST, and that K_h was a major factor affecting the mixing associated with both types of typhoon paths.

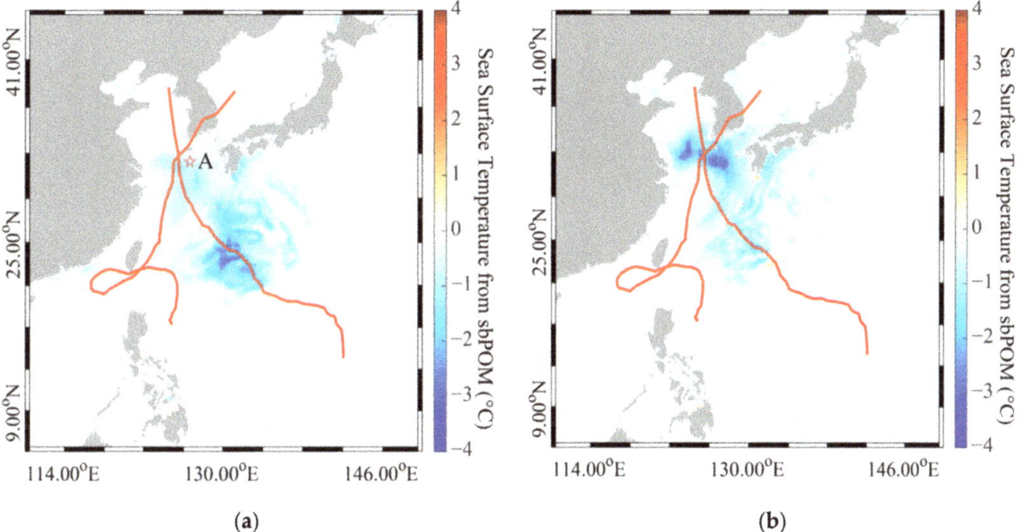

(a)   (b)

Figure 9. *Cont.*

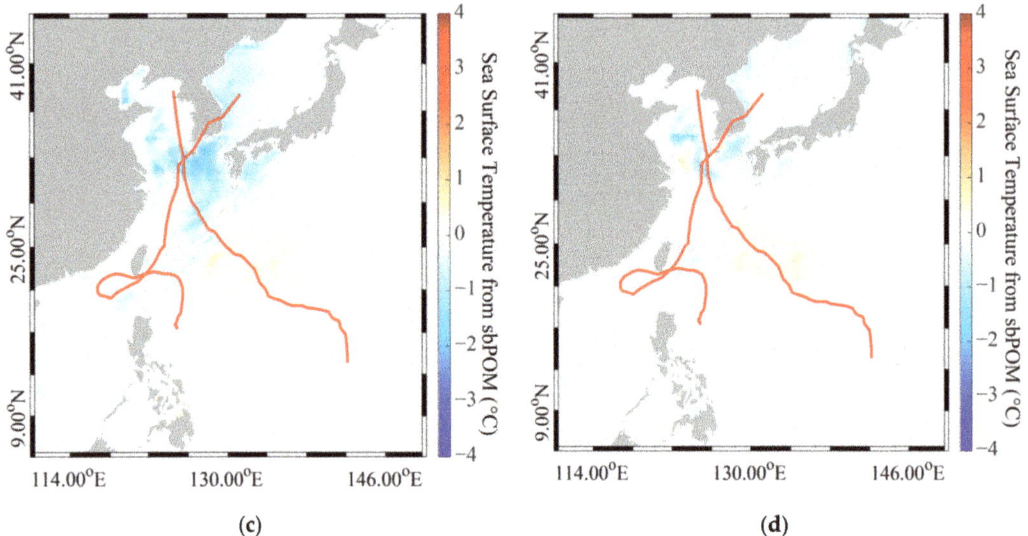

(c)

(d)

Figure 9. Daily average SSTs simulated using the sbPOM during Typhoons Tembin and Bolaven (26–30 August 2012). (**a**) Results from 26 July minus results from 27 July; (**b**) results from 27 July minus results from 28 July; (**c**) results from 28 July minus results from 29 July; (**d**) results from 29 July minus results from 30 July.

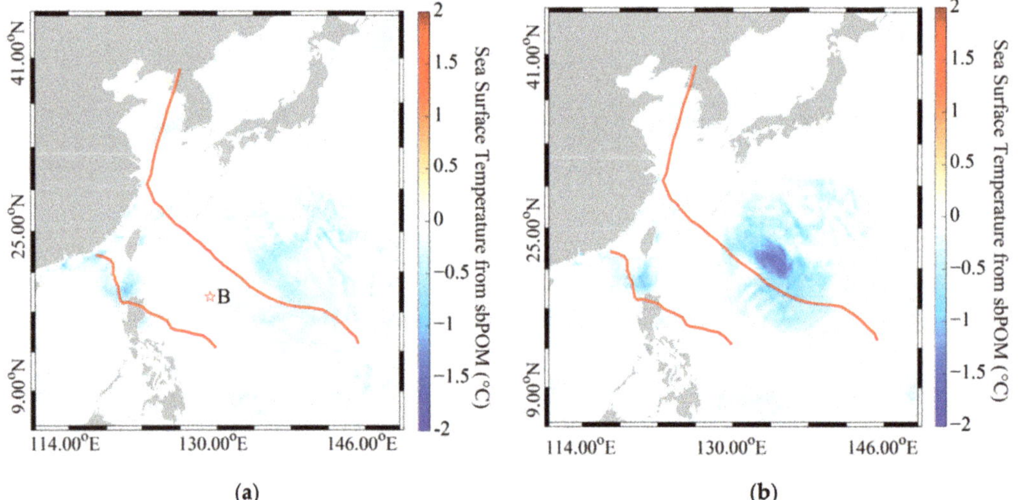

(a)

(b)

Figure 10. *Cont.*

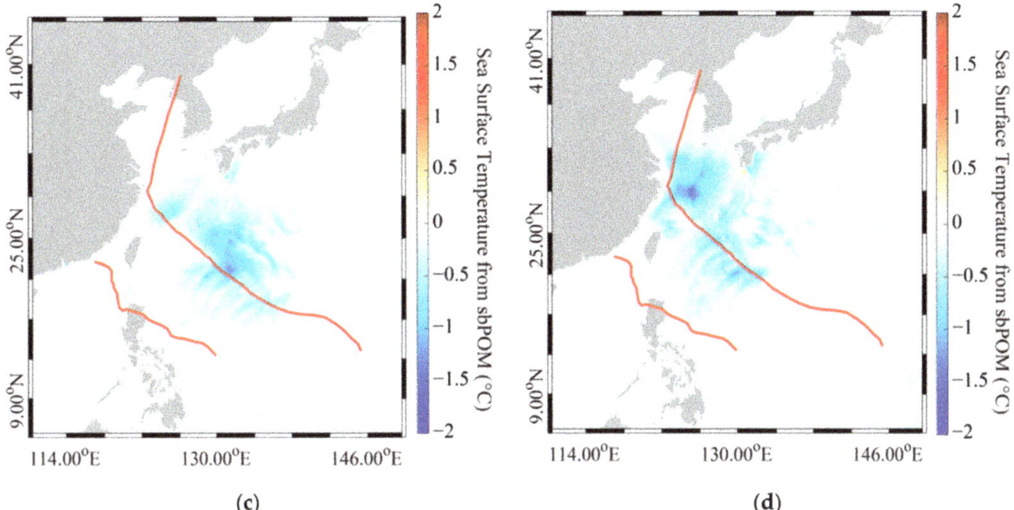

(c)   (d)

Figure 10. Daily average SSTs simulated using the sbPOM during the Chan-hom and Linfa typhoons on 7–11 July 2015. (**a**) Results from 7 July minus results from 8 July; (**b**) results from 8 July minus results from 9 July; (**c**) results from 9 July minus results from 10 July; (**d**) results from 10 July minus results from 11 July.

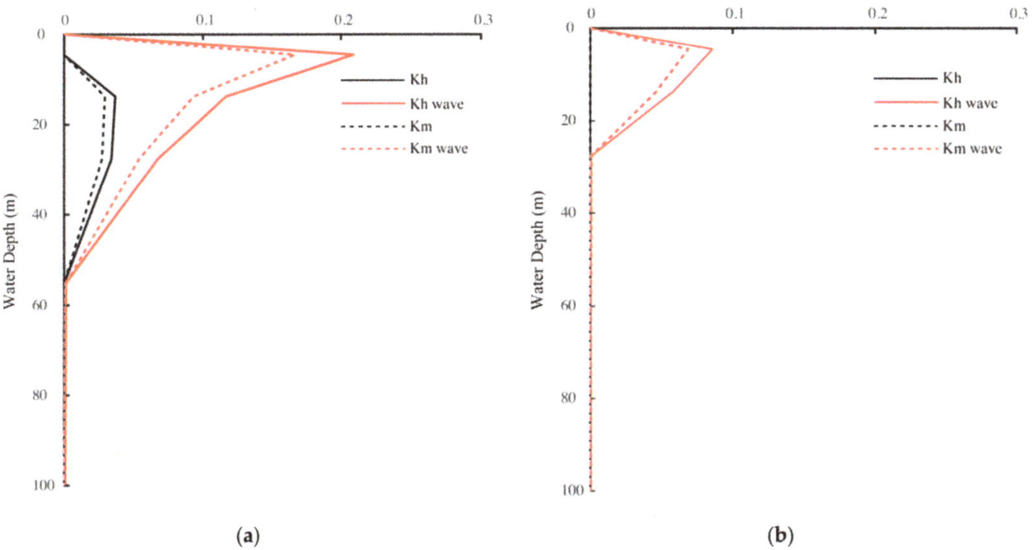

(a)   (b)

Figure 11. (**a**) The vertical mixing profile on 27 and 28 August 2012 at Site A in Figure 10a, and (**b**) the vertical mixing profile on 10 and 11 July 2015 at Site B in (**a**). Note that K_h represents the mixing induced by heat flux, and K_m represents the mixing induced by momentum.

Two sites (A and B) were selected to further investigate the vertical profile of temperature cooling (Figure 12). At Site B (Figure 12b), which is near the point where the two typhoon paths crossed, the depth of the typhoon-induced disturbance reached 150 m. The vertical temperature profile simulated by sbPOM for the parallel-type typhoon path indicated that the maximum depth of the typhoon-induced disturbance was about 50 m at

Site A (Figure 12a), and that the wave-induced effects of water temperature weakened with increasing depth. Thus, temperature reduction and typhoon-induced disturbance depth were both greater for cross-type typhoon paths than for parallel-type typhoon paths. A previous one-dimensional ocean mixed layer model (OMLM-Noh) [55] showed that the mixed layer deepened to 45 and 25 m for the typhoons Hagibis and Mitag, respectively. Our work may have identified a deeper mixed layer because the one-dimensional mixed layer model only considers mixing effects, whereas our model included Stokes drift effects.

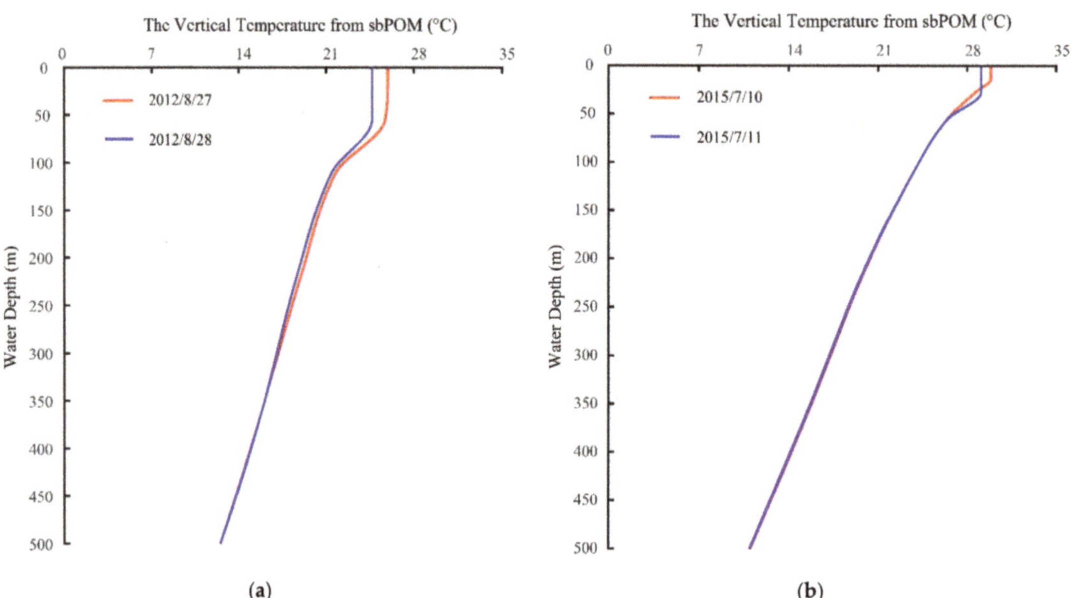

Figure 12. (a) The vertical temperature profile simulated using the sbPOM on 27 and 28 August 2012 at Site A in Figure 10a, and (b) the vertical temperature profile simulated using the sbPOM on 10 and 11 July 2015 at Site B in Figure 11a.

4. Summary and Conclusions

The aim of this work was to investigate SST cooling in response to different typhoon paths (i.e., parallel-type and cross-type) using two marine models: a wave model (WW3) and a circulation model (sbPOM). Previous studies [7,8,56] have considered the effects of typhoon waves, such as nonbreaking waves and radiation stress, on SST cooling individually. Here, we simulated SST using an sbPOM that included four effects of typhoon waves simulated by the WW3 model (breaking waves, nonbreaking waves, radiation stress, and Stokes drift), which were stronger than those at regular sea sites. We also investigated the horizontal and vertical distributions of SST cooling.

Composite H-E winds, which combine cyclonic winds using parametric Holland and ECMWF reanalysis data, were treated as the forcing field in the WW3 model. The WW3-simulated SWHs were validated against the measurements of Jason-2 altimeter, and an RMSE of less than 0.6 m and a COR of about 0.9, indicating that WW3-simulated waves were suitable for use in this study. The typhoon-wave-induced effects were calculated based on several parameters, such as SWH, mean wave period, wavelength, and dominant wave propagation velocity. When the four effects induced by typhoon waves were considered, the bias in the SST results obtained via the sbPOM simulation was improved by about 0.5 °C as compared to simulations based on Argos measurements. As noted in a previous study [46], nonbreaking waves reduce SST for individual typhoons. However, the effects of Stokes drift also have an important influence on SST cooling during binary typhoons.

Therefore, we concluded that typhoon-wave-induced effects should be included in SST simulations during typhoons, because typhoon waves influence the air–sea boundary layer (via breaking waves and Strokes drift), as well as the mixing layer (via nonbreaking waves and radiation stress).

The daily average SST distributions during the four typhoons were analyzed. We identified a finger pattern of SST cooling during both parallel-type and cross-type typhoons. SST was reduced up to 2 °C for parallel-type typhoons and up to 4 °C for cross-type typhoons. Mixing was significantly enhanced when wave-induced effects were considered; the mixing induced by heat flux was stronger than that induced by momentum. In addition, the mixing associated with cross-type typhoons was greater than that associated with parallel-type typhoons. Vertical SST profiles during the four typhoons were also studied. The results suggest that the typhoon-induced disturbance depth was 100 m for cross-type typhoons, which was deeper than the disturbance depth associated with parallel-type typhoons (50 m).

In future studies, we aim to consider sea-surface roughness and the air–sea energy exchange, including variations in the drag coefficient, using the same numeric models (WW3 and sbPOM) under binary typhoon conditions.

Author Contributions: Conceptualization, W.S. and Z.S.; methodology, W.S., Z.S. and W.Y.; validation, Z.S., W.Y. and J.L.; formal analysis, W.S. and Z.S.; investigation, Z.S.; resources, W.S.; writing—original draft preparation, Z.S. and W.S.; writing—review and editing, W.S.; visualization, W.Y. and J.L.; funding acquisition, W.S. All authors have read and agreed to the published version of the manuscript.

Funding: This research was funded by the National Key Research and Development Program of China under contract nos. 2017YFA0604901 and 2017YFA0604904, the National Natural Science Foundation of China under contract nos. 41806005 and 42076238, the National Social Science Foundation of China (Major Program) contract no. 15ZDB17 and the Science and Technology Project of Zhoushan City, China, under contract no. 2019C21008.

Institutional Review Board Statement: Not applicable.

Informed Consent Statement: Not applicable.

Data Availability Statement: Due to the nature of this research, participants of this study did not agree to share the data publicly, so supporting data are not available.

Acknowledgments: We are truly thankful for the National Centers for Environmental Prediction (NCEP) of the National Oceanic and Atmospheric Administration (NOAA) providing the source for the WAVEWATCH-III (WW3) model. The original code of Stony Brook Parallel Ocean Model (sbPOM) is available via http://www.ccpo.odu.edu (accessed on 3 June 2021) The European Centre for Medium-Range Weather Forecasts (ECMWF) provides wind data via http://www.ecmwf.int (accessed on 3 June 2021). General Bathymetry Chart of the Oceans (GEBCO) data are downloaded via ftp.edcftp.cr.usgs.gov. The Simple Ocean Data Assimilation (SODA) data are collected via https://climatedataguide.ucar.edu (accessed on 3 June 2021). The NCEP wind field and heat flux is collected via http://www.cdc.noaa.gov (accessed on 3 June 2021). The measurements from altimeter Jason-2 and Argos are accessed via https://data.nodc.noaa.gov (accessed on 3 June 2021) and http://www.argodatamgt.org (accessed on 3 June 2021), respectively.

Conflicts of Interest: The authors declare no conflict of interest.

Appendix A

The basic wave propagation balance equations in the WAVEWATCH-III (WW3) model can be briefly described as follows:

$$\frac{D}{Dt}(N(k, \theta; x, t)) = \frac{S(k, \theta; x, t)}{\sigma} \text{ and} \tag{A1}$$

$$S = S_{in} + S_{nl} + S_{ds} + S_{bot} + S_{db}, \tag{A2}$$

where the wavenumber-direction spectrum N is the basic spectrum of WW3 in terms of wavenumber k, wave propagation direction θ, space dimension x, and time dimension t; N is the wave action density spectrum; σ is the intrinsic frequency; and S(k,θ;x,t) describes the net of sources and sinks from the wavenumber-direction spectrum. The sink term S(k,θ;x,t) contains the impacts of linear and nonlinear wave propagation energy, including wind energy input S_{in}, a wave–wave interaction term S_{nl}, dissipation S_{ds}, and the empirical parameterizations of wave–bottom friction S_{bot} and depth-induced breaking S_{db}. The parameterizations of these terms are conveniently provided by the WW3 model, as described in the technical manual. The four wave-induced effects were calculated based on several parameters. Specifically, the parametrization of input/dissipation terms is referred to switch ST2+STAB2 [13,57], and the packages for processing nonlinear terms on quadruple wave–wave interactions is referred to switch DIA [5,58].

Theoretical expression of a breaking wave:

The energy dissipation rate due to wave breaking in a unit area of water column can be expressed as

$$R_{ds} = \rho_w g \int S_{ds}(\mathbf{k})d\mathbf{k}, \tag{A3}$$

where ρ_w is the density of sea water, g is the gravity constant (9.8 m/s^2), and R_{ds} is the downward input of turbulent kinetic energy flux due to wave breaking on the sea surface. In practice, a unified analytical form of wave breaking energy dissipation is derived as follows [59]:

$$R_{dis} = 2.97\gamma\rho_w g\beta_*^{-2}\omega_p E, \tag{A4}$$

where

$$\beta_* = \frac{g}{u_*\omega_p}, \tag{A5}$$

$$E = \int F(\mathbf{k})d\mathbf{k}, \tag{A6}$$

$$\omega_p = \frac{2\pi}{T_p}, \tag{A7}$$

where γ is energy dissipated per unit of white crown accounting for the percentage of total wave energy dissipation, generally γ = 0.1; β_* is the wave age; E is spectral density; **k** is wave number vector; F(**k**) is the two-dimensional wave spectrum; ω_p is the spectral peak angular frequency; u_* is the friction velocity (=$\sqrt{c_d U_{10}}$); c_d is the drag coefficient; and T_p is the dominant wave period.

Theoretical expression of a nonbreaking wave:

The parameter schemes of vertical diffusion coefficient K_h and vertical eddy viscosity coefficient K_m are

$$K_m = \frac{2ak^2\lambda}{\pi T}e^{\frac{2\pi z}{\lambda}}, \tag{A8}$$

$$K_h = \frac{2Pk^2}{g}\delta\beta^3 W^3 e^{\frac{gz}{\beta^2 W^2}}, \tag{A9}$$

where k is the Kaman constant; a is the amplitude, which is twice the effective significant wave height; T is the fluctuation period; λ is the wavelength; z is the distance from the sea surface to a certain depth; β_* is the wave age; P is a dimensionless variable related to the Richardson number coefficient, and δ is wave steepness. Generally, at the sea surface k = 0.4, β = 1.0, P = 0.1, δ = 0.1, π = 3.14, and g = 9.8 m/s^2.

Theoretical expression of radiation stress:

The components of radiation stress S_{xx}, S_{yy}, and S_{xy} are calculated as follows:

$$S_{xx} = kE\left(\frac{k_x k_x}{k_x^2 + k_y^2}F_{CS}F_{CC} - F_{SC}F_{SS}\right) + E_D, \tag{A10}$$

$$S_{yy} = kE\left(\frac{k_y k_y}{k_x^2 + k_y^2}F_{CS}F_{CC} - F_{SC}F_{SS}\right) + E_D, \tag{A11}$$

$$S_{xy} = S_{yx} = \sqrt{k_x^2 + k_y^2}E\frac{k_x k_y}{k^2}F_{CS}F_{CC}, \tag{A12}$$

where

$$F_{SC} = \frac{\sin hk(z+h)}{\cos hkD}, F_{CC} = \frac{\cos hk(z+h)}{\cos hkD}, \tag{A13}$$

$$F_{SS} = \frac{\sin hk(z+h)}{\sin hkD}, F_{CS} = \frac{\cos hk(z+h)}{\sin hkD}, \tag{A14}$$

$$E = \frac{1}{16}\rho_w gH_s^2, \tag{A15}$$

where ρ_w is the seawater density; H_s is the significant wave height; $g = 9.8$ m/s^2; k_x and k_y are the wave numbers in the x and y dimensions, respectively; $D = H + \eta$, where h is the seabed topography and η is the sea surface undulation; and $\int_{-h}^{\eta+} E_D dz = E/2$ when $z \neq \eta$ and $E_d = 0$.

Theoretical Expression of Stokes drift:

The Stokes drift of a single-frequency deep-water gravity wave can be expressed as

$$U_s = U_{ss}e^{\frac{8\pi^2 z}{gT^2}}k \tag{A16}$$

$$U_{ss} = \frac{2\pi^3 H_s^2}{gT^3}, \tag{A17}$$

where U_s is the Stokes drift rate on the ocean surface, k is the unit wavenumber vector of the fluctuation, H_s is significant wave height, T is mean wave period, $g = 9.8$ m/s^2, and z is the water depth (z = 0 at the sea surface; z > 0 above the water surface).

Appendix B

The Stony Brook Parallel Ocean Model (sbPOM) follows the principles of the POM. The advantage of the sbPOM is that computational efficiency is improved due to its use of a parallel computing environment. However, because the σ coordinate system is used in the vertical direction in the sbPOM, z coordinates must be converted into σ coordinates. This conversion is performed as follows:

$$\sigma = \frac{z - \eta}{H + \eta}, \tag{A18}$$

where H(x,y) is the bottom terrain in the horizontal x and y dimensions; η(x,y,t) is the sea level fluctuation in horizontal dimensions x and y at time dimension t, which is integrated from the bottom (z = $-$H) to the sea surface (z = η); and σ is set from -1 to 0. Using this equation, the basic equations of the sbPOM under the σ-coordinate system can be expressed as

$$\frac{\partial uD}{\partial t} + \frac{\partial u^2 D}{\partial x} + \frac{\partial uvD}{\partial y} + \frac{\partial u\omega}{\partial \sigma} - fvD + gD\frac{\partial \eta}{\partial x} + \frac{gD^2}{\rho_0}\int_\sigma^0\left[\frac{\partial \rho}{\partial x} - \frac{\sigma}{D}\frac{\partial D}{\partial x}\frac{\partial \rho}{\partial \sigma}\right]d\sigma = \frac{\partial}{\partial \sigma}\left[\frac{K_M}{D}\frac{\partial u}{\partial \sigma}\right] + F_x \tag{A19}$$

$$\frac{\partial vD}{\partial t} + \frac{\partial uvD}{\partial x} + \frac{\partial v^2 D}{\partial y} + \frac{\partial v\omega}{\partial \sigma} + fuD + gD\frac{\partial \eta}{\partial y} + \frac{gD^2}{\rho_0}\int_\sigma^0\left[\frac{\partial \rho}{\partial y} - \frac{\sigma}{D}\frac{\partial D}{\partial y}\frac{\partial \rho}{\partial \sigma}\right]d\sigma = \frac{\partial}{\partial \sigma}\left[\frac{K_M}{D}\frac{\partial v}{\partial \sigma}\right] + F_y \tag{A20}$$

where

$$D = H + \eta \tag{A21}$$

and where u, v, and ω represent the velocities under the respective σ-coordinates; ρ is the mean fluctuation value; g is the gravitational acceleration; K_M is the vertical viscosity coefficient; and F_x and F_y are the horizontal viscosity terms. The use of these equations should reduce the truncation errors associated with the calculation of the pressure gradient term in an σ-coordinate system over steep topography.

References

1. Pun, I.F.; Lin, I.I.; Lo, M.H. Recent increase in high tropical cyclone heat potential area in the Western North Pacific Ocean. *Geophys. Res. Lett.* **2013**, *40*, 4680–4684. [CrossRef]
2. Emanuel, K.A. An air-sea interaction theory for tropical cyclones. Part I: Steady-state maintenance. *J. Atmos. Sci.* **1985**, *43*, 585–605. [CrossRef]
3. Rastigejev, Y.; Suslov, S.A. Effect of evaporating sea spray on heat fluxes in a marine atmospheric boundary layer. *J. Phys. Oceanogr.* **2019**, *49*, 1927–1948. [CrossRef]
4. Jun, K.C.; Jeong, W.M.; Choi, J.Y.; Park, K.S.; Jung, K.T.; Kim, M.W.; Chae, J.W.; Qiao, F.L. Simulation of the extreme waves generated by Typhoon Bolaven (1215) in the East China Sea and Yellow Sea. *Acta Oceanol. Sin.* **2015**, *34*, 19–28. [CrossRef]
5. Shao, W.Z.; Sheng, Y.X.; Li, H.; Shi, J.; Ji, Q.Y.; Tai, W.; Zuo, J.C. Analysis of wave distribution simulated by WAVEWATCH-III model in typhoons passing Beibu Gulf, China. *Atmosphere* **2018**, *8*, 265. [CrossRef]
6. Xie, B.; Zhang, F. Impacts of typhoon track and island topography on the heavy rainfalls in Taiwan associated with Morakot (2009). *Mon. Weather Rev.* **2011**, *140*, 3379–3394. [CrossRef]
7. Price, J.F. Upper ocean response to a hurricane. *J. Phys. Oceanogr.* **1981**, *11*, 153–175. [CrossRef]
8. Bender, M.A.; Ginis, I.; Kurihara, Y. Numerical simulations of tropical cyclone-ocean interaction with a high-resolution coupled model. *J. Geophys. Res.* **1993**, *98*, 23245–23263. [CrossRef]
9. Doong, D.J.; Tsai, C.H.; Chen, Y.C.; Peng, J.P.; Huang, C.J. Statistical analysis on the long-term observations of typhoon waves in the Taiwan sea. *J. Mar. Sci. Eng.* **2015**, *23*, 893–900.
10. Li, X.F. The first sentinel-1 SAR image of a typhoon. *Acta Oceanol. Sin.* **2015**, *34*, 1–2. [CrossRef]
11. Wang, X.; Cheng, G. Tracking typhoon-generated swell in the western North Pacific Ocean using satellite altimetry. *Chin. J. Oceanol. Limn.* **2015**, *33*, 1157–1163. [CrossRef]
12. Monaldo, F.M.; Thompson, D.R.; Pichel, W.G.; Clemente-Colón, P. A systematic comparison of QuikSCAT and SAR ocean surface wind speeds. *IEEE Trans. Geosci. Remote Sens.* **2004**, *42*, 283–291. [CrossRef]
13. Bao, L.; Peng, G.; Peng, H.; Jia, Y.; Qi, G. First accuracy assessment of the HY-2A altimeter sea surface height observations: Cross-calibration results. *Adv. Space Res.* **2015**, *55*, 90–105. [CrossRef]
14. Wamdi, T. The WAM model—A third generation ocean wave prediction model. *J. Phys. Oceanogr.* **1988**, *18*, 1775–1810.
15. Sheng, Y.X.; Shao, W.Z.; Li, S.Q.; Zhang, Y.M.; Yang, H.W.; Zuo, J.C. Evaluation of typhoon waves simulated by WaveWatch-III model in shallow waters around Zhoushan Islands. *J. Ocean U. China* **2019**, *18*, 365–375. [CrossRef]
16. Rogers, W.E.; Hwang, P.A.; Wang, D.W. Investigation of wave growth and decay in the SWAN model: Three regional-scale applications. *J. Phys. Oceanogr.* **2003**, *33*, 366–389. [CrossRef]
17. Zheng, K.W.; Sun, J.; Guan, C.L.; Shao, W.Z. Analysis of the global swell and wind-sea energy distribution using WAVEWATCH III. *Adv. Meteorol.* **2016**, *7*, 1–9. [CrossRef]
18. Hu, Y.Y.; Shao, W.Z.; Shi, J.; Sun, J.; Ji, Q.Y.; Cai, L.N. Analysis of the typhoon wave distribution simulated in WAVEWATCH-III model in the context of Kuroshio and wind-induced current. *J. Oceanol. Limn.* **2020**, *38*, 1692–1710. [CrossRef]
19. Yang, Z.H.; Shao, W.Z.; Ding, Y.Y.; Shi, J.; Ji, Q.Y. Wave simulation by the SWAN model and FVCOM considering the sea-water level around the Zhoushan islands. *J. Mar. Sci. Eng.* **2020**, *8*, 783. [CrossRef]
20. Shao, W.Z.; Hu, Y.Y.; Zheng, G.; Cai, L.N.; Zou, J.C. Sea state parameters retrieval from cross-polarization Gaofen-3 SAR data. *Adv. Space Res.* **2019**, *65*, 1025–1034. [CrossRef]
21. Shao, W.Z.; Jiang, X.W.; Nunziata, F.; Marino, A.; Corcione, V. Analysis of waves observed by synthetic aperture radar across ocean fronts. *Ocean Dynam.* **2020**, *70*, 1–11. [CrossRef]
22. Shao, W.Z.; Ding, Y.Y.; Li, J.C.; Gou, S.P.; Nunziata, F.; Yuan, X.Z.; Zhao, L.B. Wave retrieval under typhoon conditions using a machine learning method applied to Gaofen-3 SAR imagery. *Can. J. Remote Sens.* **2019**, *45*, 723–732. [CrossRef]
23. Jiang, X.P.; Zhong, Z.; Liu, C.X. The effect of typhoon-induced SST cooling on typhoon intensity: The case of Typhoon Chanchu (2006). *Adv. Atmos. Sci.* **2008**, *25*, 1062–1072. [CrossRef]
24. Oke, P.R.; Schiller, A. Impact of Argo, SST, and altimeter data on an eddy-resolving ocean reanalysis. *Geophys. Res. Lett.* **2007**, *341*, L19601. [CrossRef]
25. Tsai, Y.; Chern, C.S.; Wang, J. The upper ocean response to a moving typhoon. *J. Oceanogr.* **2008**, *64*, 115–130. [CrossRef]
26. Shay, L.K.; Black, P.G.; Mariano, A.J.; Hawkins, J.D.; Elsberry, R.L. Upper ocean response to hurricane Gilbert. *J. Geophys. Res.* **1992**, *97*, 20227–20248. [CrossRef]
27. Sheng, J.; Zhai, X.M.; Greatbatch, R.J. Numerical study of the storm-induced circulation on the Scotian Shelf during Hurricane Juan using a nested-grid ocean model. *Prog. Oceanogr.* **2006**, *70*, 233–254. [CrossRef]
28. Sakaida, F.; Kawamura, H.; Toba, Y. Sea surface cooling caused by typhoons in the Tohoku area in August 1989. *J. Geophys. Res.* **1998**, *103*, 1053–1065. [CrossRef]
29. Lin, I.I.; Liu, W.T.; Wu, C.C.; Chiang, J.; Sui, C.H. Satellite observations of modulation of surface winds by typhoon-induced upper ocean cooling. *Geophys. Res. Lett.* **2003**, *30*, 1131. [CrossRef]

30. Pun, I.F.; Lin, I.I.; Lien, C.C.; Wu, C.C. Influence of the size of supertyphoon Megi (2010) on SST cooling. *Mon. Weather Rev.* **2018**, *146*, 661–677. [CrossRef]
31. Guan, S.D.; Zhao, W.; Huthnance, J.; Tian, J.W.; Wang, J. Observed upper ocean response to Typhoon Megi (2010) in the northern South China Sea. *J. Geophys. Res.* **2014**, *119*, 3134–3157. [CrossRef]
32. Chan, J.C.L.; Duan, Y.; Shay, L.K. Tropical cyclone intensity change from a simple ocean-atmosphere coupled model. *J. Atmos. Sci.* **2001**, *58*, 154–172. [CrossRef]
33. Lee, C.Y.; Chen, S.S. Stable boundary layer and its impact on tropical cyclone structure in a coupled atmosphere—Ocean model. *Mon. Weather Rev.* **2014**, *142*, 1927–1944. [CrossRef]
34. Wu, C.C.; Tu, W.T.; Pun, I.F.; Lin, I.I.; Peng, M.S. Tropical cyclone-ocean interaction in Typhoon Megi (2010)—A synergy study based on ITOP observations and atmosphere-ocean coupled model simulations. *J. Geophys. Res.* **2016**, *121*, 153–167. [CrossRef]
35. Zhu, T.; Zhang, D.L. The impact of the storm-induced SST cooling on hurricane intensity. *Adv. Atmos. Sci.* **2006**, *23*, 14–22. [CrossRef]
36. Cione, J.J.; Uhlhorn, E.W. Sea surface temperature variability in hurricanes: Implications with respect to intensity change. *Mon. Weather Rev.* **2003**, *128*, 1783–1796. [CrossRef]
37. Weatherford, C.L.; Gray, W.M. Typhoon structure as revealed by aircraft reconnaissance. Part I: Data analysis and climatology. *Mon. Weather Rev.* **1988**, *116*, 1032–1043. [CrossRef]
38. Allahdadi, N. Numerical Experiments of Hurricane Impact on Vertical Mixing and De-Stratification of the Louisiana Shelf Waters. Doctoral Dissertation, Louisiana State University, Baton Rouge, LA, USA, 2014.
39. Allahdadi, M.N.; Li, C.Y. Numerical Simulation of Louisiana Shelf Circulation under Hurricane Katrina. *J. Coast. Res.* **2018**, *34*, 67–80. [CrossRef]
40. Allahdadi, M.N.; Li, C.Y. Effect of stratification on current hydrodynamics over Louisiana shelf during Hurricane Katrina. *Water Sci. Eng.* **2017**, *10*, 154–165. [CrossRef]
41. Xian, Z.; Chen, K. Numerical analysis on the effects of binary interaction between typhoons Tembin and Bolaven in 2012. *Adv. Meteorol.* **2019**, *4*, 1–16. [CrossRef]
42. Price, J.M.; Reed, M.; Howard, M.K.; Johnson, W.R.; Ji, Z.G.; Marshall, C.F.; Guinasso, N.L.; Rainey, G.B. Preliminary assessment of an oil-spill trajectory model using satellite-tracked, oil-spill-simulating drifters. *Environ. Model. Softw.* **2006**, *21*, 258–270. [CrossRef]
43. Nittis, K.; Perivoliotis, L.; Korres, G.; Tziavos, C.; Thanos, I. Operational monitoring and forecasting for marine environmental applications in the Aegean Sea. *Environ. Modell. Softw.* **2006**, *21*, 243–257. [CrossRef]
44. Jordi, A.; Wang, D.P. Sbpom: A parallel implementation of Princenton ocean model. *Environ. Model. Softw.* **2012**, *38*, 59–61. [CrossRef]
45. Lv, X.; Yuan, D.; Ma, X.; Tao, J. Wave characteristics analysis in Bohai Sea based on ECMWF wind field. *Ocean Eng.* **2014**, *91*, 159–171. [CrossRef]
46. Zhou, L.M.; Li, Z.B.; Mu, L.; Wang, A.F. Numerical simulation of wave field in the South China Sea using WAVEWATCH III. *Chin. J. Oceanol. Limn.* **2014**, *37*, 656–664. [CrossRef]
47. Stopa, J.E.; Cheung, K.F. Intercomparison of wind and wave data from the ECMWF reanalysis interim and the NECP climate forecast system reanalysis. *Ocean Model.* **2014**, *75*, 65–83. [CrossRef]
48. Roemmich, D.; Gilson, J. The 2004–2008 mean and annual cycle of temperature, salinity, and steric height in the global ocean from the Argo program. *Prog. Oceanogr.* **2009**, *82*, 81–100. [CrossRef]
49. Jiang, M.; Ke, X.; Liu, Y.; Lei, W. Estimating the sea state bias of Jason-2 altimeter from crossover differences by using a three-dimensional nonparametric model. *IEEE J. Sel. Top. Appl. Earth Observ. Remote Sens.* **2016**, *9*, 5023–5043. [CrossRef]
50. Aijaz, S.; Ghantous, M.; Babanin, A.V.; Ginis, I.; Thomas, B.; Wake, G. Nonbreaking wave-induced mixing in upper ocean during tropical cyclones using coupled hurricane-ocean-wave modeling. *J. Geophys. Res.* **2017**, *122*, 3939–3963. [CrossRef]
51. Seroka, G.; Miles, T.; Xu, Y.; Kohut, J.; Schofield, O.; Glenn, S. Hurricane Irene sensitivity to stratified coastal ocean cooling. *Mon. Weather Rev.* **2016**, *144*, 3507–3530. [CrossRef]
52. Saji, N.H.; Xie, S.P.; Tam, C.Y. Satellite observations of intense intraseasonal cooling events in the tropical south Indian Ocean. *Geophys. Res. Lett.* **2006**, *33*, 70–84. [CrossRef]
53. Wu, X.; Fei, J.F.; Huang, X.G.; Cheng, X.P.; Ren, J.Q. Statistical classification and characteristics analysis of binary tropical cyclones over the western North Pacific Ocean. *J. Trop. Meteorol.* **2011**, *17*, 335–344.
54. Nelson, N.B. The wake of Hurricane Felix. *Int. J. Remote Sens.* **1996**, *17*, 2893–2895. [CrossRef]
55. Yang, Y.J.; Sun, L.; Duan, A.M.; Li, Y.B.; Fu, Y.F.; Yan, T.F.; Wang, Z.Q.; Xian, T. Impacts of binary typhoons on upper ocean environments in November 2007. *J. Appl. Remote Sens.* **2012**, *6*, 3583–3596. [CrossRef]
56. Zhu, P.; Wang, Y.; Chen, S.S.; Curcic, M.; Gao, C. Impact of storm-induced cooling of sea surface temperature on large turbulent eddies and vertical turbulent transport in the atmospheric boundary layer of hurricane Isaac. *J. Geophys. Res.* **2016**, *121*, 861–876. [CrossRef]
57. Tolman, H.L.; Chalikov, D. Source Terms in a Third-Generation Wind Wave Model. *J. Phys. Oceanogr.* **1996**, *26*, 2497–2518. [CrossRef]

58. Hasselmann, S.; Hasselmann, K.; Allender, J.H.; Barnett, T.P. Computations and parameterizations of the nonlinear energy transfer in agravity-wave spectrum, Part II: Parameterizations of the nonlinear energy transfer for application in wave models. *J. Phys. Oceanogr.* **1985**, *15*, 1378–1391. [CrossRef]
59. Guan, C.L.; Hu, W.; Sun, J.; Li, R.L. The whitecap coverage model from breaking dissipation parametrizations of wind waves. *J. Geophys. Res.* **2007**, *112*, C05013. [CrossRef]

Journal of
Marine Science and Engineering

MDPI

Article

Effect of Depth-Induced Breaking on Wind Wave Simulations in Shallow Nearshore Waters off Northern Taiwan during the Passage of Two Super Typhoons

Shih-Chun Hsiao [1], Han-Lun Wu [1], Wei-Bo Chen [2,*], Wen-Dar Guo [2], Chih-Hsin Chang [2] and Wen-Ray Su [2]

[1] Department of Hydraulic and Ocean Engineering, National Cheng Kung University, Tainan City 70101, Taiwan; schsiao@mail.ncku.edu.tw (S.-C.H.); hlwu627@mail.ncku.edu.tw (H.-L.W.)
[2] National Science and Technology Center for Disaster Reduction, New Taipei City 23143, Taiwan; wdguo@ncdr.nat.gov.tw (W.-D.G.); chang.c.h@ncdr.nat.gov.tw (C.-H.C.); wrsu@ncdr.nat.gov.tw (W.-R.S.)
* Correspondence: wbchen@ncdr.nat.gov.tw; Tel.: +886-2-8195-8611

Citation: Hsiao, S.-C.; Wu, H.-L.; Chen, W.-B.; Guo, W.-D.; Chang, C.-H.; Su, W.-R. Effect of Depth-Induced Breaking on Wind Wave Simulations in Shallow Nearshore Waters off Northern Taiwan during the Passage of Two Super Typhoons. *J. Mar. Sci. Eng.* **2021**, *9*, 706. https://doi.org/10.3390/jmse9070706

Academic Editor: Christos Stefanakos

Received: 20 April 2021
Accepted: 24 June 2021
Published: 26 June 2021

Abstract: Super Typhoons Maria (2018) and Lekima (2019) were adopted for this case study, although they only passed the northern offshore waters of Taiwan without making landfall. A direct modification technique was employed to create the atmospheric conditions for a wave-circulation model to hindcast large typhoon-driven waves. The radius of the modified scale (R_{trs}) for a hybrid typhoon wind plays an important role in the significant wave height (SWH) simulations during the passage of typhoons. The maximum increment in peak SWH reached 3.0 m and 5.0 m in the deep ocean for Super Typhoons Maria (2018) and Lekima (2019), respectively if the R_{trs} was increased from $4 \times R_{max}$ (radius of the maximum wind) to $7 \times R_{max}$. The SWHs induced by the typhoon winds in the surf zone were more sensitive to different wave-breaking formulations used in the wave-circulation model. The maximum difference in peak SWH reached 2.5 m and 1.2 m for Super Typhoons Maria (2018) and Lekima (2019), respectively, when the wave-breaking formulations of BJ78 (proposed by Battjes and Janssen in 1978) and CT93 (proposed by Church and Thornton in 1993) were introduced to the wave-circulation model. The SWH simulations in the surf zone were insensitive to the wave-breaking criterion (γ) during the passage of typhoons. In shallow nearshore waters, the utilization of a constant γ for the wave-circulation model always produces peak SWHs that are smaller than those using γ based on local steepness or peak steepness.

Keywords: depth-induced wave breaking; wave-breaking formulation; wave-breaking criterion; shallow nearshore waters

1. Introduction

Ocean surface wind waves are a dominant process in coastal, nearshore and offshore regions worldwide. Understanding the characteristics of cyclone-driven extreme waves, their variability and historical and projected future changes are important considerations for the sustainable development of coastal and offshore infrastructures and the management of coastal resources and ecosystems. The energy in ocean surface waves is transmitted from the wind. The wind patterns above the ocean surface have been affected as the upper ocean has warmed, consequently resulting in stronger ocean waves. According to the report from Reguero et al. [1], the energy of ocean waves, i.e., wave height, has grown over the past seven decades, which could have significant implications for coastal communities and ecosystems. Since the contributions of ocean surface waves to extreme total water levels are substantial at open coasts, they have mostly been considered in many local studies [2–5].

Predicting wave heights accurately in coastal and nearshore areas is essential for a number of human activities there, such as renewable energy applications, aquaculture, maritime transport and infrastructure; furthermore, information on both swells and wind seas is important to coastal applications. For example, the prediction of locally generated wind seas is essential for high-speed passenger ferries, whereas information about the propagation of low-frequency swells in coastal areas and fjords is critical for the design of coastal structures or in the planning of marine operations.

Depth-induced wave breaking is one of the most dominant hydrodynamic processes occurring in coastal regions. This wave breaking not only controls the amount of wave energy impacting our coastlines and coastal defenses but also plays a crucial role in driving many nearshore processes, such as sediment transport, bottom morphology [6] and turbulence, which has been shown to be very important for local ecology [7]. Wave breaking also induces radiation stresses that drive wave-induced setup and currents [8], both of which are important for coastal engineering design and management. However, despite the importance and relevance toward our knowledge of wave hydrodynamics, depth-induced wave breaking is still poorly understood, which is partially due to its highly nonlinear nature; therefore, it is heavily parameterized in most wind wave models.

Many parameterizations for the wave-breaking index have been proposed and reported in previous studies. These parameterizations include dependencies of the wave-breaking index on the offshore wave steepness [9,10], a dissipation rate based on a normalized surf zone width [11], the offshore wave height and the inverse Iribarren number [12]. Ruessink et al. [13] also introduced a parameterization of the wave-breaking index that linearly increases with the local nondimensional depth based on the peak period.

This study provides a critical and objective assessment of three specialized depth-induced wave-breaking models and wave-breaking criteria, which are based on state-of-the-art breaking wave formulations. The aim of this paper is to study the effect of the depth-induced wave-breaking formulation and wave-breaking criteria on the hindcasts of SWH in shallow nearshore waters off northern Taiwan during the passage of Super Typhoons Maria in 2018 and Lekima in 2019, as well as to better understand which hindcast is more influential in wave height hindcasting inside the surf zone. The details of the materials and methods are presented in the following section, and the results of model validation and a series of designed model experiments are described in Section 3. A discussion and uncertainties of the present study are given in Section 4. Finally, Section 5 summarizes and concludes the paper.

2. Materials and Methods

2.1. Description of the Study Cases

Super Typhoon Maria was a powerful tropical cyclone that affected Guam (the United States), the Ryukyu Islands (Japan), Taiwan and East China in early July 2018. Maria became a tropical storm and passed the Mariana Islands on July 4 and rapidly intensified the next day due to favorable environmental conditions. Maria reached its first peak intensity on 6 July, and a second stronger peak intensity with 1-min sustained winds of 270 km/h (equivalent to category 5 super typhoon status on the Saffir-Simpson scale) and a minimum pressure of 915 hPa was reached on 9 July. Maria finally made landfall over the Fujian Province, China, early on 11 July after crossing the Yaeyama Islands and passing the northern offshore waters of Taiwan on 10 July. Figure 1a demonstrates the track and arrival times of Super Typhoon Maria in 2018.

Super Typhoon Lekima originated from a tropical depression that was developed in the eastern Philippines on 30 July 2019. Lekima became a tropical storm and was named on 4 August. Under favorable environmental conditions, Lekima intensified and reached its peak with 1-min sustained winds of 250 km/h (equivalent to a category 4 super typhoon status on the Saffir-Simpson scale) and a minimum pressure of 925 hPa on 8 August. Lekima made landfall in the Zhejiang Province, China, on late 9 August and made its second landfall in the Shandong Province, China, on 11 August after moving across eastern China. The track and arrival times of Super Typhoon Lekima are illustrated in Figure 1b.

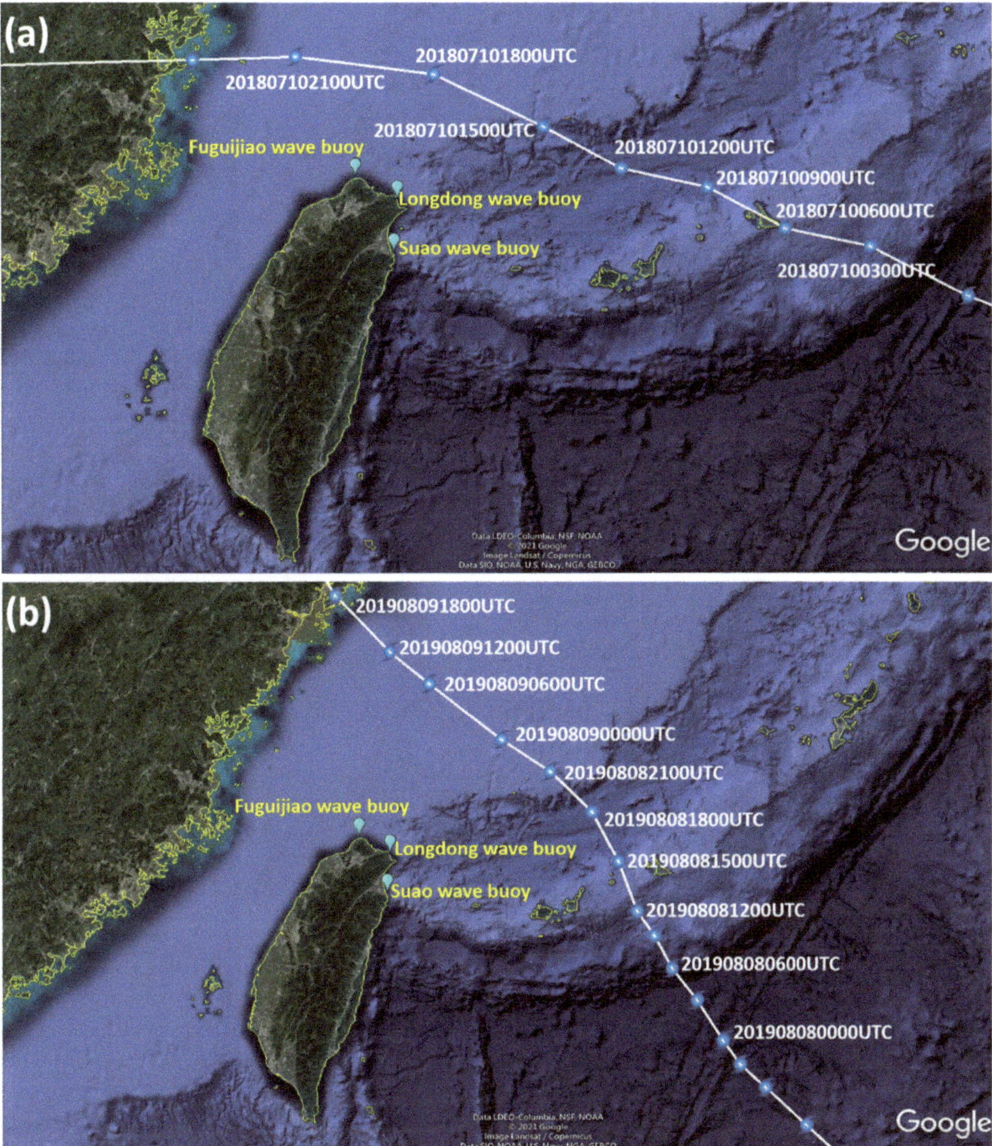

Figure 1. Tracks and arrival times of (**a**) Super Typhoon Maria in 2018 and (**b**) Lekima in 2019. The locations of wave buoys adopted in the present study are marked in cyan (Data source: Regional Specialized Meteorological Center (RSMC) Tokyo-Typhoon Center).

2.2. Information on Offshore Wave Buoys

Three wave buoys, namely, Fuguijiao, Longdong and Suao, located in the northern and northeastern offshore waters of Taiwan, were selected for model validation because they are closest to the tracks of Super Typhoons Maria in 2018 and Lekima in 2019 (as shown in Figure 1). The sampling frequency of wave buoys is 2 Hz for 10 min at the beginning of each hour with an accuracy of ± 10 cm for the SWH measurements according to the annual buoy observation data report from the CWB. Information on the coordinates of the three wave buoys and their corresponding water depths is listed in Table 1.

Table 1. Information on wave buoys.

Buoy Name	Longitude (°E)	Latitude (°N)	Water Depth (m)
Fuguijiao	121.5336	25.3036	30
Longdong	121.9225	25.0978	27
Suao	121.8758	24.6247	23

Data source: The Central Weather Bureau and Water Resource Agency of Taiwan.

2.3. Direct Modification of Typhoon Winds from ERA5

Since the mid-1960s, many parametric cyclone wind models have been proposed [14,15] and widely used to mimic the wind distribution of typhoons [16–19] because of their simplicity. Parametric cyclone wind models could be used to accurately reconstruct the wind distributions near the center of the typhoon; however, they are unable to accurately reproduce wind speeds in regions far from the center of the typhoon. In contrast, reanalysis wind data obtained from the dynamical model with data assimilation show a superior performance for hindcasting the winds outside of the typhoon's center but are generally inferior for the hindcasts of maximum typhoon wind speed [20–24]. A direct modification technique recommended by Pan et al. [20] was applied in the present study to take advantage of combining the parametric cyclone wind model and reanalysis wind data and maintain a reliable structure for the entire typhoon wind field.

$$W_{DM} = \begin{cases} W_{ERA5}\left[\frac{r}{R_{\max}}\left(\frac{W_{Bmax}}{W_{Emax}} - 1\right) + 1\right] & r < R_{\max} \\ W_{ERA5}\left[\frac{R_{trs}-r}{R_{trs}-R_{\max}}\left(\frac{W_{Bmax}}{W_{Emax}} - 1\right) + 1\right] & R_{\max} \leq r \leq R_{trs} \\ W_{ERA5} & r > R_{trs} \end{cases} \tag{1}$$

where W_{DM} is the wind speed at an arbitrary grid within the model domain through direct modification, W_{ERA5} is the wind speed extracted from ERA5 (the fifth-generation reanalysis of the European Centre for Medium-Range Weather Forecasts for the global climate and weather) at an arbitrary point in the computational grid, W_{Bmax} is the maximum wind speed of the best track typhoon issued by the Regional Specialized Meteorological Center (RSMC) Tokyo-Typhoon Center, W_{Emax} is the maximum wind speed of the typhoon among the hourly ERA5 wind fields, r is the radial distance from an arbitrary grid within the model domain to the eye of the typhoon, R_{trs} is the radius of the modified scale (also known as the radius of the transitional zone) and $R_{\max}$ is the radius at the maximum typhoon wind speed. $R_{\max}$ can be expressed as a function of W_{Bmax} and the latitude of the typhoon's center:

$$R_{\max} = m_0 + m_1 \times W_{Bmax} + m_2(\phi - 25) \tag{2}$$

where ϕ is the latitude of the typhoon's center. In Equation (2), m_0, m_1 and m_2 were set to 38.0 (in n·mi), -0.1167 (in n·mi·kt^{-1}) and -0.0040 (in n·mi$^{\circ -1}$), respectively, according to the results derived from Knaff et al. [25] for the Western Pacific typhoon basin. R_{trs} is considered to be a key factor in determining the accuracy of wind fields; therefore, various R_{trs} will be employed to create hybrid typhoon wind fields to better understand their effect on wind wave hindcasting.

2.4. Configuration of the Wave-Circulation Modeling System

A seamless cross-scale hydrodynamic model based on an unstructured grid and triangular mesh served as the ocean circulation model in the wave-circulation modeling system. The hydrodynamic model is called SCHISM (semi-implicit cross-scale hydroscience integrated system model), which has been implemented by Zhang et al. [26] and other developers around the world. Similar to the SELFE (semi-implicit Eulerian-Lagrangian finite element/volume model, the predecessor of the SCHISM developed by Zhang and Baptista [27]), the SCHISM also avoids the severest stability constraints in the numerical model by means of a highly efficient semi-implicit scheme [28]. Hence, the high-performance calculations can be performed even when a very high spatial resolution mesh is used in the SCHISM. The splitting between internal and external modes could derive a numerical error [29], which is eliminated through the no-mode-splitting technique implemented in the SCHISM. A depth-averaged (two-dimensional (2D)) ocean circulation model is sufficient for simulating typhoon-driven hydrodynamics. Additionally, fewer computing resources and execution times are required for a 2D model. Therefore, a 2D model, i.e., SCHISM-2D, is selected for wind wave modeling in the present study. The SCHISM and its predecessor SELFE are multipurpose models that have been widely applied to the simulation of hydrodynamics and water quality transportation in coastal and estuarine environments in Taiwan, e.g., evaluation of storm tide-induced coastal inundation [30,31], assessment of tidal stream energy [32,33], transport of suspended sediment and fecal coliform [34,35]. The Manning coefficient and time step were set as 0.025 and 120 s for the barotropic ocean model, respectively, according to the geological characteristics of the seafloor in the waters surrounding Taiwan and the numerical stability of the SCHISM-2D.

The WWM-III is a derivative work from the original WWM-II (wind wave model version III, developed by Roland [36]). The WWM-III is a third-generation spectral wave model that is able to simulate and predict the ocean surface sea state [37]. The wave action balance equation governing the WWM-III is solved by the fractional step method on an unstructured grid. The number of directional bins is 36 with a minimum of 0° and a maximum direction of 360°. The number of frequency bins is 36, the lowest limit of the discrete wave frequency is 0.04 Hz and a highest frequency limit of the discrete wave period is 1.0 Hz. The peak enhancement factor is specified as 3.3 for the JONSWAP (Joint North Sea Wave Project, [38]) spectra, while the wave breaking coefficients for the constant and bottom friction coefficients are 0.78 and 0.067, respectively, in WWM-III.

To enhance the information exchange efficiency between the ocean circulation and wind wave models and to eliminate the interpolation errors from the two models, SCHISM-2D and WWM-III take advantage of sharing the same subdomains and parallelization through the same domain decomposition scheme. Additionally, time steps of 120 s and 600 s were used in the SCHISM-2D and WWM-III models, respectively, to improve the computational performance of the coupled model, SCHISM-WWM-III. The fully coupled SCHISM-WWM-III modeling system has been applied to predict, simulate and hindcast typhoon-driven storm tides and waves [21–24,39], as well as long-term wave energy resources in the offshore waters of Taiwan [40–42].

For a successful storm surge, tide and wave modeling, the size of the computational domain must be large enough to accommodate the full typhoon; otherwise, the simulations would be affected by the boundary conditions [43,44]. The present study developed a computational domain spanning east longitudes from 105° to 140° and north latitudes from 15° to 31° (as shown in Figure 2). This large domain is composed of 540,510 nonoverlapping triangular elements and 276,639 unstructured grid points. A local-scale bathymetric dataset covering the area from east longitude 100° to 128° and north latitude from 4° to 29° with a spatial resolution of 200 m was provided by the Department of Land Administration and the Ministry of the Interior, Taiwan. The latest global-scale bathymetric product released from the General Bathymetric Chart of the Oceans (GEBCO), GEBCO_2020 Grid, was employed to incorporate a local-scale bathymetric dataset (as mentioned above) to construct the gridded bathymetric data in the SCHISM-WWM-III modeling system.

Eight main tidal constituents (M_2, S_2, N_2, K_2, K_1, O_1, P_1 and Q_1) extracted from a regional inverse tidal model (China Seas and Indonesia [45]) were utilized to generate the tidal elevation and horizontal velocity at the ocean boundaries of the SCHISM-WWM-III. The inverse barometric effect for tidal elevation at the boundary nodes was also considered. Since wave-generating typhoons were completely within the computational domain in the present study, open boundary conditions for the waves were not always required [42,46].

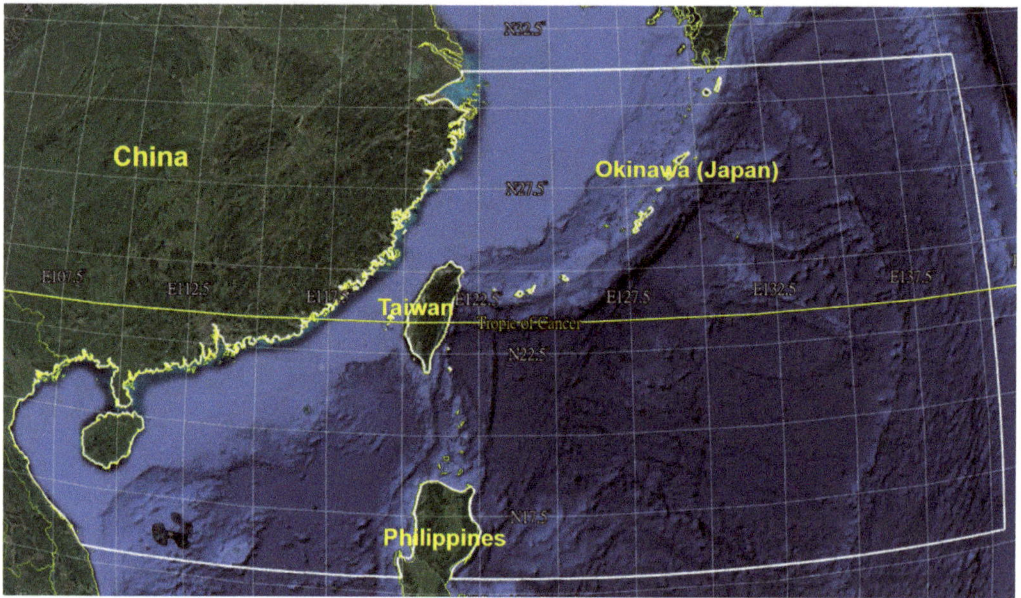

Figure 2. Coverage of the computational domain for the wave-circulation modeling system.

2.5. Parameterization of Depth-Induced Wave Breaking

Three widely used parameterizations for characterizing depth-induced wave breaking in nearshore shallow waters were adopted in the present study and are briefly reviewed in this section. Battjes and Beji [47] proposed that the total energy dissipation can be distributed over the wave spectrum in proportion to the spectral density based on extensive laboratory experiments; hence, the wave-induced wave breaking (S_{break}) in the wave action density function is calculated as follows:

$$S_{break}(\sigma,\theta) = D_{total}\frac{E(\sigma,\theta)}{E_{total}} \tag{3}$$

where $S_{break}(\sigma,\theta)$ is the wave-induced wave breaking in spectral space (σ, θ), D_{total} is the total energy dissipation rate due to depth-induced wave breaking, $E(\sigma,\theta)$ is the wave energy in spectral space (σ, θ) and E_{total} is the total wave energy.

2.5.1. BJ78 Model

In the BJ78 model (proposed by Battjes and Janssen [48]), the total energy dissipation due to depth-induced wave breaking, D_{tot_BJ78} is given as follows:

$$D_{tot\_BJ78} = -\frac{1}{4}\alpha_{BJ}Q_b\bar{f}H_{max}^2 \tag{4}$$

where α_{BJ} is the proportionality parameter, Q_b represents the fraction of breaking waves, $\bar{f}$ is the averaged frequency and H_{max} represents the maximum individual wave height and is defined as a proportion of the water depth in the local area (d):

$$H_{max} = \gamma d \tag{5}$$

where γ is the wave-breaking index. The fraction of breaking waves, Q_b is determined by the Rayleigh distribution truncated at an upper limit with H_{max} (maximum wave height). Therefore, the fraction of breaking waves is implicitly expressed as follows:

$$\frac{1 - Q_b}{-\ln Q_b} = \left(\frac{H_{rms}}{H_{max}}\right)^2 \tag{6}$$

where H_{rms} is the root-mean-square wave height.

2.5.2. TG83 Model

The TG83 model can be considered a variant of BJ78, and the total energy dissipation due to depth-induced wave breaking in the TG83 model (proposed by Thornton and Guza [49]), D_{tot_TG83} is formulated as follows:

$$D_{tot\_TG83} = -\frac{B^3\,\overline{f}}{4\,d}\int_0^\infty H^3 p_b(H)dH \tag{7}$$

where B is the proportionality coefficient, H is the wave height and $p_b(H)$ is the fraction of breaking waves at each wave height. In which, $p_b(H)$ can be given as follows:

$$p_b(H) = W(H)p(H) \tag{8}$$

where $p(H)$ is the Rayleigh wave height probability density function and can be expressed as follows:

$$p(H) = \frac{2H}{H_{rms}^2}\exp\left[-\left(\frac{H}{H_{rms}}\right)^2\right] \tag{9}$$

The weighting function, $W(H)$ is defined as in the TG83 model:

$$W(H) = \left(\frac{H_{rms}}{\gamma_{TG}d}\right)^n \tag{10}$$

where the calibration parameter $n = 4$, γ_{TG} is the wave-breaking index for the TG83 model and is set to 0.42 according to Thornton and Guza [49,50].

2.5.3. CT93 Model

The total energy dissipation due to the depth-induced wave breaking implemented in the CT93 model (proposed by Church and Thornton [51]) is related to H_{rms} through periodic linear bore theory, which can be given as follows:

$$D_{tot\_CT93} = -\frac{3\sqrt{\pi}}{16}\overline{f}B^3\frac{H_{rms}^3}{d}M\left[1 - \frac{1}{\left[1 + (H_{rms}/\gamma d)^2\right]^{5/2}}\right] \tag{11}$$

in which M is expressed as follows:

$$M = 1 + \tanh\left[8\left(\frac{H_{rms}}{\gamma} - 1.0\right)\right] \tag{12}$$

2.6. Parameterization of the Wave-Breaking Index

2.6.1. Constant Wave Breaking Criterion (γ)

The parameter γ, given in Equation (4), is an adjustable coefficient that allows for the effects of the bottom slope on the waves and is one of the important parameters in the energy dissipation formulation. Miche [52] derived the theoretical value of $\gamma = 0.88$ for a flat bottom. Based on the average of several reasonable wave breaking observations, Battjes and Janssen [48] suggested a constant breaker index of $\gamma = 0.8$. The values for the breaking index γ range from 0.60 to 0.83; however, $\gamma = 0.73$ has been used in the most operational third-generation wave models, which is taken as the averaged value over a larger data set [9]. In the present study, a value of $\gamma = 0.78$ was specified for the SWH simulation following the results from Chen et al. [21] and Hsiao et al. [22–24] if the wave-circulation model adopted a constant breaker index.

2.6.2. Wave Breaking Criterion Based on Local Steepness or Peak Steepness

The wave breaking process in shallow nearshore waters is affected by the seafloor profile (i.e., bottom slope) and incident wave steepness. Battjes and Stive [9] found that a hyperbolic tangent function could be used for predicting the SWHs in shallow-water coastal areas and a relationship between the breaking index γ and the wave steepness s_0:

$$\gamma = 0.5 + 0.4\text{tanh}(33s_0) \tag{13}$$

in which s_0 can be expressed as follows:

$$s_0 = H_{rms0}/\lambda_0 \tag{14}$$

where H_{rms0} is the root-mean-square wave height in deep waters, λ_0 is the wavelength and is defined as $\lambda_0 = \sqrt{gd}/f$, f is the wave period and g is the acceleration of gravity. The wave breaking criterion in Equation (13) is regarded as the local steepness or peak steepness based on f.

3. Results

The wind fields during Super Typhoon Maria from 1–15 July in 2018 and Super Typhoon Lekima from 1–15 August in 2019 were extracted from the ERA5 reanalysis and were corrected by the direct modification method expressed in Equation (1). The hybrid typhoon winds derived from Equation (1) with various R_{trs} values were imposed in the SCHISM-WWM-III modeling system to compare the performance of the resulting storm wave hindcast. The effect of the wave-breaking formulation and wave-breaking criterion on the simulation of the SWH in the shallow nearshore waters off northern Taiwan during the passage of Super Typhoons Maria and Lekima was investigated by conducting several designed model experiments, as listed in Table 2.

Table 2. Designed numerical experiments in the present study.

Scenario	Wave-Breaking Formulation	Wave-Breaking Criterion	R_{trs}
S_NO1	BJ87	$\gamma = 0.78$	$4R_{max}$
S_NO2	TG83	$\gamma_{GT} = 0.42$	$4R_{max}$
S_NO3	CT93	$\gamma = 0.78$	$4R_{max}$
S_NO4	BJ87	Based on Local Steepness	$4R_{max}$
S_NO5	BJ87	Based on Peak Steepness	$4R_{max}$

3.1. Validation for Typhoon-Driven SWHs with Various R_{trs}

The inverse distance weighting method was employed to convert the hourly ERA5 and hybrid typhoon winds from the structured grid (at a horizontal resolution of 31 km) to the unstructured grid for the SCHISM-WWM-III modeling system. Comparisons of the SWH time series between model hindcasts using the different radii of the modified scale (R_{trs}) and the corresponding measurements are depicted in Figure 3a–c for the Fuguijiao (Figure 3a), Longdong (Figure 3b) and Suao (Figure 3c) buoys during the passage of Super Typhoon Maria in 2018. The hindcasted peak wave heights are underestimated by 1.5 m and 2.0 m for the Fuguijiao and Longdong wave buoys, respectively, when the original ERA5 winds are used in the SCHISM-WWM-III modeling system but slightly overestimate the peak wave height within 0.5 m for the Suao buoy. Modified ERA5 winds with various R_{trs} were imposed on the SCHISM-WWM-III modeling system to improve the performance of typhoon wave hindcasts. The hybrid winds with $R_{trs} = 3R_{max}$, $R_{trs} = 4R_{max}$, $R_{trs} = 5R_{max}$, $R_{trs} = 6R_{max}$ and $R_{trs} = 7R_{max}$ are called the H_3R$_{max}$, H_4R$_{max}$, H_6R$_{max}$, H_6R$_{max}$ and H_7R$_{max}$ winds, respectively. As shown in Figure 3a–c, the hindcasted peak wave heights for all three wave buoys increased when using a larger R_{trs} value. For instance, the hindcasted peak wave height was raised to 7.5 m for the Fuguijiao wave buoy by exerting H_4R$_{max}$ winds on the SCHISM-WWM-III modeling system (as shown in Figure 3a). Similar phenomena can be found in Figure 3b (for the Longdong wave buoy) and Figure 3c (for the Suao wave buoy). The hindcasted peak wave heights were always underestimated for the Fuguijiao wave buoy even though H_7R$_{max}$ winds were utilized. However, the SCHISM-WWM-III modeling system overestimated the peak wave height for the Suao wave buoy once the hybrid typhoon winds were used for the meteorological boundary conditions. Figure 4 illustrates the spatial distribution of the difference in the maximum hindcasted SWH between adopting the winds from H_4R$_{max}$ and the original ERA5 for Super Typhoon Maria in 2018. Significant differences can be detected along the track of Super Typhoon Maria, and the extents with differences exceeding 3.0 m occurred in the deep ocean. The spatial distributions of the difference in maximum hindcasted SWH between employing the winds from H_5R$_{max}$ and H_4R$_{max}$, H_6R$_{max}$ and H_4R$_{max}$ and H_7R$_{max}$ and H_4R$_{max}$ are demonstrated in Figure 5a–c, respectively. The difference in maximum hindcasted SWH using different winds increased when R_{max} was enlarged. The maximal differences were always distributed over the right side of Super Typhoon Maria in 2018, where the wind speed was highest (as shown in Figure 5a–c). The same validation process was conducted to verify the SWH hindcasts for Super Typhoon Lekima in 2019. Figure 6 presents the comparisons of the SWH time series between model hindcasts using the different R_{trs} values and the corresponding measurements or the Fuguijiao (Figure 6a), Longdong (Figure 6b) and Suao (Figure 6c) buoys during the period of Super Typhoon Lekima in 2019. The hindcasts from the use of the H_3R$_{max}$ and H_4R$_{max}$ winds are more satisfactory for all three wave buoys. The improvements in the hindcasted peak SWH for Super Typhoon Lekima in 2019 are obvious; for example, the difference in the maximum hindcasted SWH between inputting the winds from H_4R$_{max}$ and the original ERA5 were up to 5 m in the deep ocean (as shown in Figure 7). The spatial distributions of the difference in maximum hindcasted SWH between applying the winds from H_5R$_{max}$ and H_4R$_{max}$, H_6R$_{max}$ and H_4R$_{max}$ and H_7R$_{max}$ and H_4R$_{max}$ for Super Typhoon Lekima in 2019 are shown in Figure 8a,c, respectively. Similar to the hindcasts of Super Typhoon Maria in 2018, maximal differences were also detected on the right side of Super Typhoon Lekima in 2019, where the wind speed was strongest and increased with the increase in R_{trs}. According to the resulting storm wave hindcasts of Super Typhoons Maria in 2018 and Lekima in 2019, the hybrid wind field with $R_{trs} = 4R_{max}$, i.e., H_4R$_{max}$ winds, was adopted as the atmospheric forcing for the SCHISM-WWM-III modeling system to conduct a series of numerical experiments.

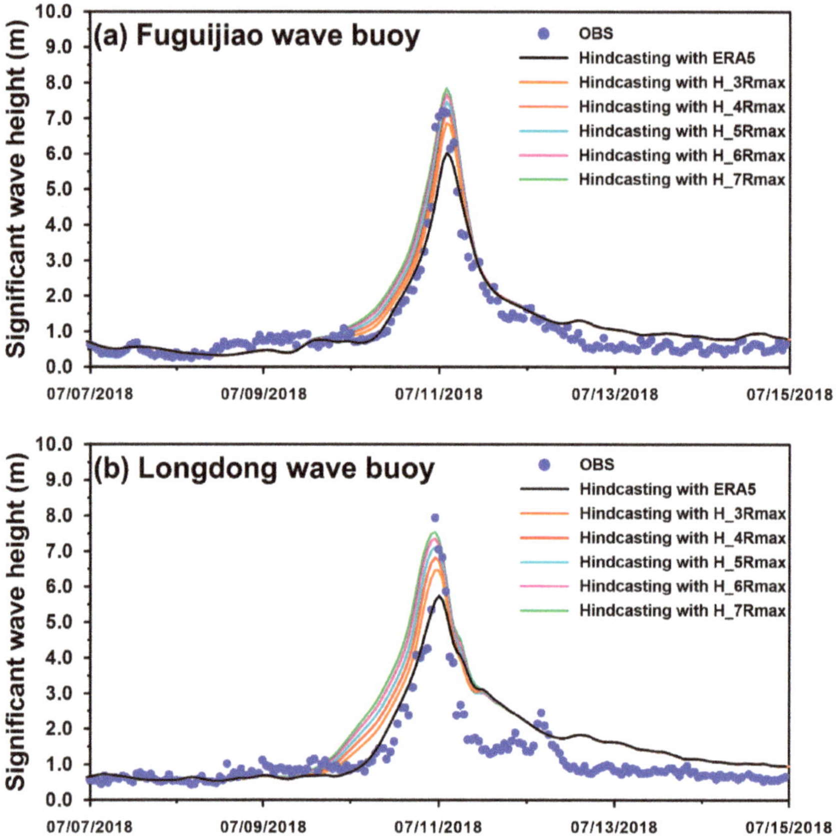

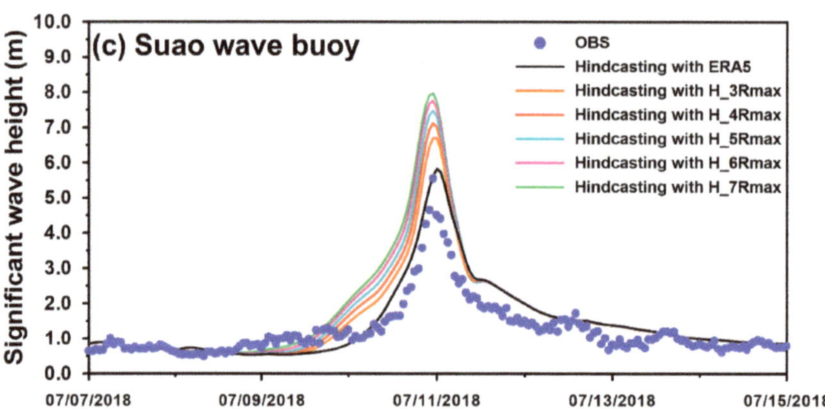

Figure 3. *Cont.*

Figure 3. Comparison of the time series of the SWH between the wave simulation that used the winds from the original ERA5, from the hybrid winds with various R_{trs} and the corresponding observations for the (**a**) Fuguijiao, (**b**) Longdong and (**c**) Suao wave buoys during the passage of Super Typhoon Maria in 2018.

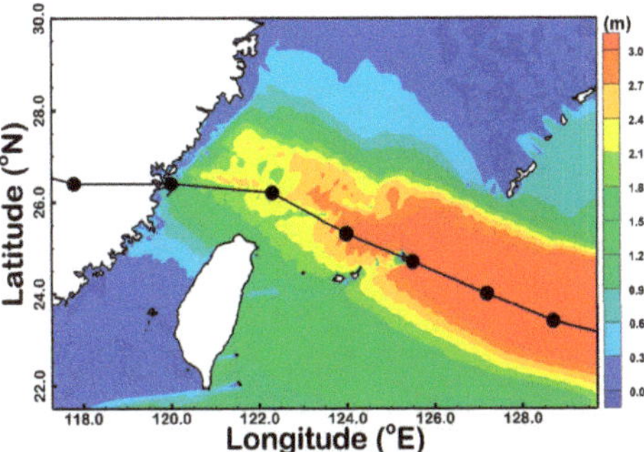

Figure 4. Spatial distribution of the difference in maximum SWH using the winds from H_4R$_{max}$ and the original ERA5 during the passage of Super Typhoon Maria in 2018.

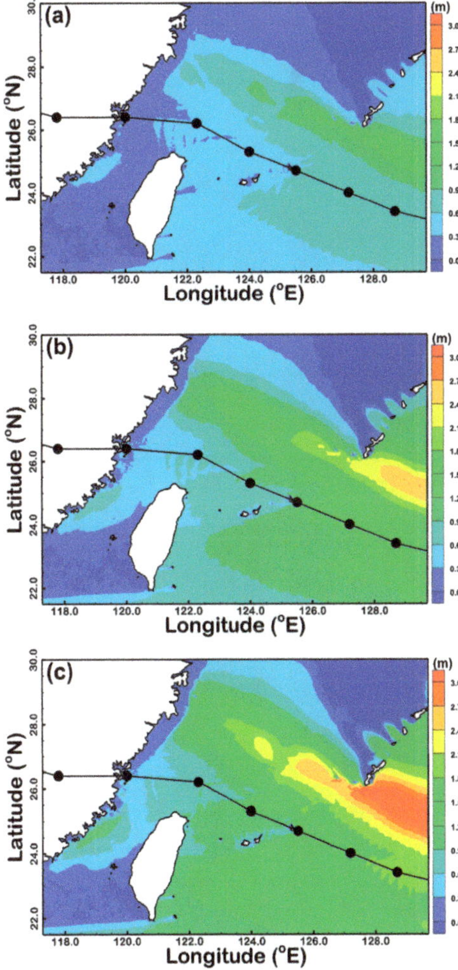

Figure 5. Spatial distribution of the difference in maximum SWH using the winds from (**a**) H_5R$_{max}$ and H_4R$_{max}$, (**b**) H_6R$_{max}$ and H_4R$_{max}$ and (**c**) H_7R$_{max}$ and H_4R$_{max}$ during the passage of Super Typhoon Maria in 2018.

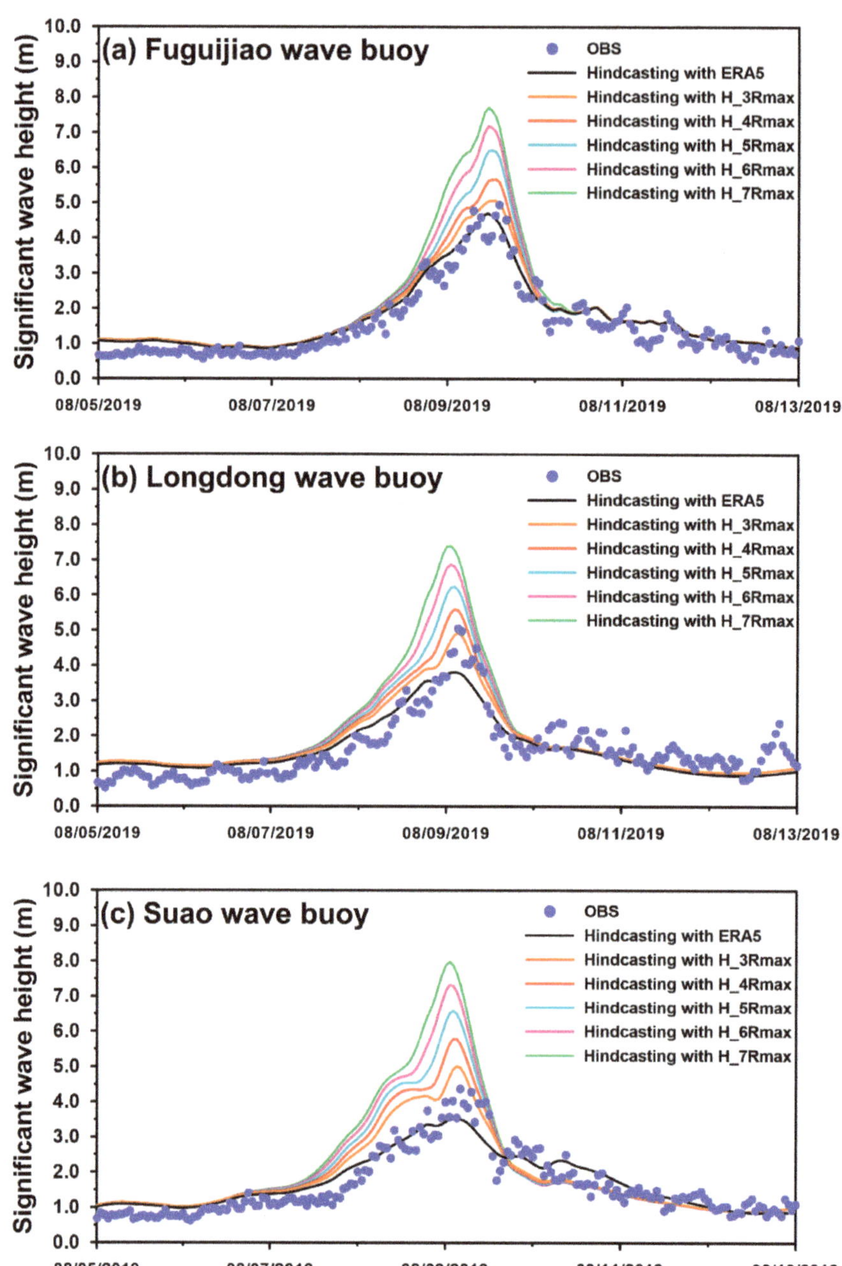

Figure 6. Comparison of the time series of the SWH between the wave simulation that used the winds from the original ERA5, from the hybrid winds with various R_{trs}, and the corresponding observations for the (**a**) Fuguijiao, (**b**) Longdong and (**c**) Suao wave buoys during the passage of Super Typhoon Lekima in 2019.

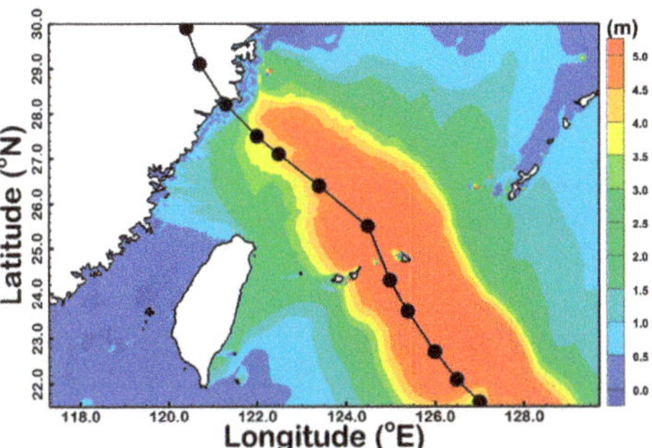

Figure 7. Spatial distribution of the difference in maximum SWH using the winds from H_4R$_{max}$ and the original ERA5 during the passage of Super Typhoon Lekima in 2019.

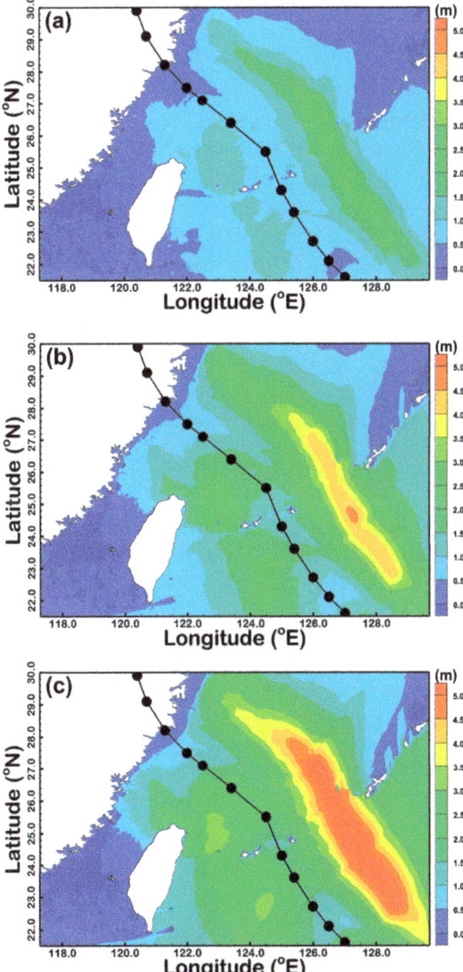

Figure 8. Spatial distribution of the difference in maximum SWH using the winds from (**a**) H_5R$_{max}$ and H_4R$_{max}$, (**b**) H_6R$_{max}$ and H_4R$_{max}$ and (**c**) H_7R$_{max}$ and H_4R$_{max}$ during the passage of Super Typhoon Lekima in 2019.

3.2. Effect of the Wave-Breaking Formulation on the Typhoon-Driven SWH Simulation in the Surf Zone

To assess the effects of different wave-breaking formulations on the wave hydrodynamics in shallow nearshore waters, three depth-induced wave-breaking parameterizations, introduced in Section 2.5, were applied to hindcast the typhoon waves with the constant wave breaking criterion (γ) and same wind forcing ($R_{trs} = 4R_{max}$ winds). As listed in Table 2, the designed numerical experiments refer to S_NO1 and S_NO3 for the BJ87 and CT93 wave-breaking models, respectively, with a constant γ of 0.78. A wave-breaking index of 0.42, namely, γ_{GT}, is specified for the TG83 model in the present study, which is labeled S_NO2 in Table 2. The spatial distribution of the difference in the maximum SWH between the scenarios of S_NO1 and S_NO2 and S_NO1 and S_NO3 during passage Super Typhoon Maria in 2018 are depicted in Figure 9. Figure 9a,b illustrate the differences in S_NO1 and S_NO2 and S_NO1 and S_NO2, respectively. Large storm waves were generated in the deep ocean and subsequently dissipated due to the decrease in water depth across a surf zone toward the shore. Although the surf zone usually lies in a shallow area where the water depth ranges from 5 to 10 m below sea level, nearshore areas with water depths greater than −20 m are shown in Figure 9 to effectively distinguish among the difference in hindcasted SWHs by different wave-breaking models. The surf zones along the north and northeast coasts of Taiwan are quite narrow because these areas are characterized by a very steep sloping seafloor. The differences in the maximum SWH between the S_NO1 and S_NO3 scenarios (Figure 9b) are more significant than those between the S_NO1 and S_NO2 (Figure 9a) scenarios. Two points, namely, P1 and P2, located in the north and northeastern shallow nearshore waters, were selected to compare the time series of hindcasted SWHs for the S_NO1, S_NO2 and S_NO3 scenarios. The comparison results are presented in Figure 10a for P1 and Figure 10b for P2. The differences between the three scenarios can only be found during the passage of Super Typhoon Maria in 2018, i.e., from midday on 10 July to midday on 11 July in 2018. The maximal difference in the hindcasted SWH is approximately 2.5 m between the S_NO1 and S_NO3 scenarios for P2 (as shown in Figure 10b). However, the maximal difference in hindcasted SWH is within 1.0 m between the S_NO2 and S_NO3 scenarios for both P1 and P2 (as shown in Figure 10a,b). Similar results are also shown in Figure 11 (spatial distribution) and Figure 12 (time series comparison) for Super Typhoon Lekima in 2019. The maximal differences in hindcasted SWH in the shallow nearshore waters were smaller for Super Typhoon Lekima in 2019 than those for Super Typhoon Maria in 2018. The maximal difference in hindcasted SWH is approximately 1.5 m at P2 during the passage of Super Typhoon Lekima in 2019 (as shown in Figure 12b).

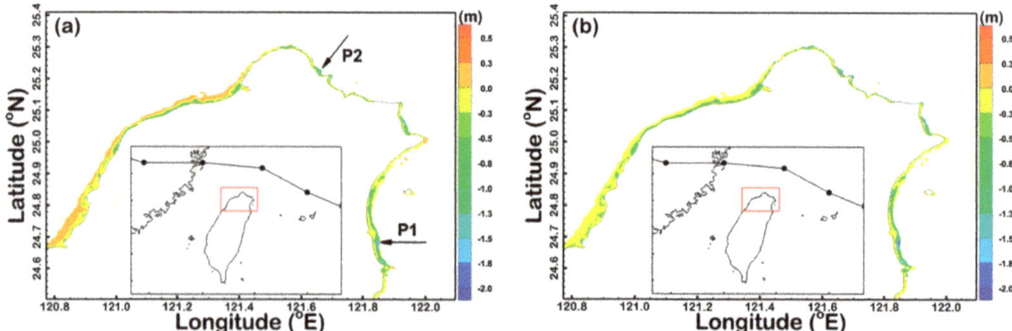

Figure 9. Spatial distribution of the difference in maximum SWH between the scenarios of (**a**) S_NO1 and S_NO2 and (**b**) S_NO1 and S_NO3 during the passage of Super Typhoon Maria in 2018. The areas with water depths greater than −20 m are shown.

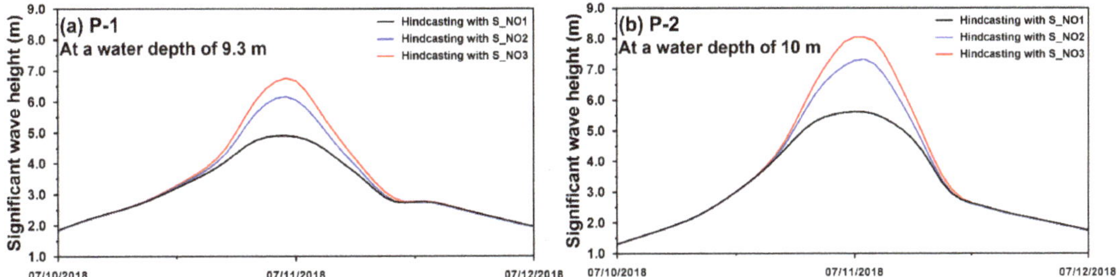

Figure 10. Time series of hindcasted SWHs for (**a**) P1 and (**b**) P2 using the S_NO1, S_NO2 and S_NO3 scenarios during the passage of Super Typhoon Maria in 2018.

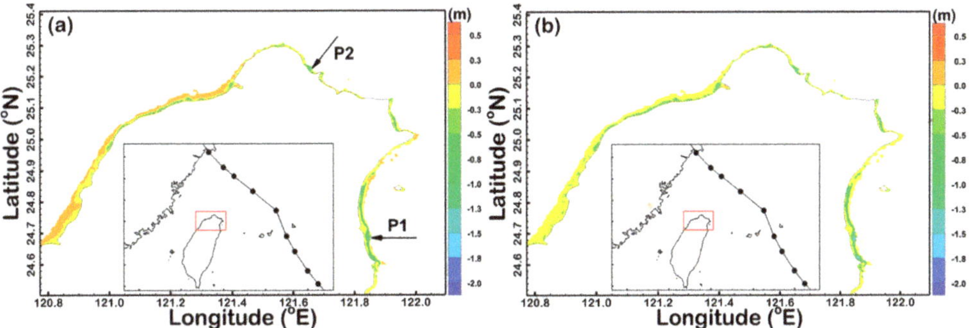

Figure 11. Spatial distribution of the difference in the maximum SWH between the scenarios of (**a**) S_NO1 and S_NO2 and (**b**) S_NO1 and S_NO3 during the passage of Super Typhoon Lekima in 2019. The areas with water depths greater than −20 m are shown.

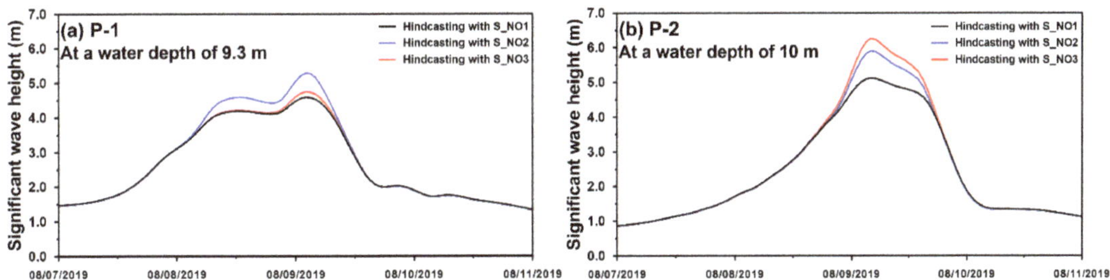

Figure 12. Time series of hindcasted SWHs for (**a**) P1 and (**b**) P2 using the S_NO1, S_NO2 and S_NO3 scenarios during the passage of Super Typhoon Lekima in 2019.

3.3. Effect of the Wave-Breaking Criterion on the Typhoon-Driven SWH Simulation in the Surf Zone

Many parameterizations of the wave-breaking criterion have been implemented within wind wave spectral models to yield substantial improvements in model hindcasts, simulations and forecasts. Hence, three scenarios, called S_NO1, S_NO4 and S_NO5, exploiting the BJ87 wave-breaking model with $R_{trs} = 4R_{max}$ wind forcing and different wave-breaking criteria (γ), were applied to hindcast storm waves in the shallow nearshore waters of northern Taiwan during the passage of Super Typhoon Maria in 2018 and Lekima in 2019. The spatial distributions of the difference in the maximum SWH hindcasted by S_NO1 and S_NO4 and S_NO1 and S_NO5 during the passage of Super Typhoon Maria in 2018 are illustrated in Figure 13a,b, respectively. The differences caused by different wave-breaking criteria γ are smaller than those caused by different wave-breaking formulations. Figure 14a,b demonstrate the time series of hindcasted SWHs for P1 and P2 using the S_NO1, S_NO4 and S_NO5 scenarios during the passage of Super Typhoon Maria in 2018. The maximal difference is within 1.0 m at P2 (as shown in Figure 14b) but is less than 0.5 m at P1 (Figure 14a). The same phenomena were also detected in both the spatial distribution (as shown in Figure 15a,b) and time series of the hindcasted SWHs (as shown in Figure 16a,b) during the passage of Super Typhoon Lekima in 2019. Interestingly, the SWHs hindcasted by a constant γ (S_NO1 scenario) are usually smaller than those hindcasted by γ based on local steepness (S_NO4 scenario) and peak steepness (S_NO5 scenario).

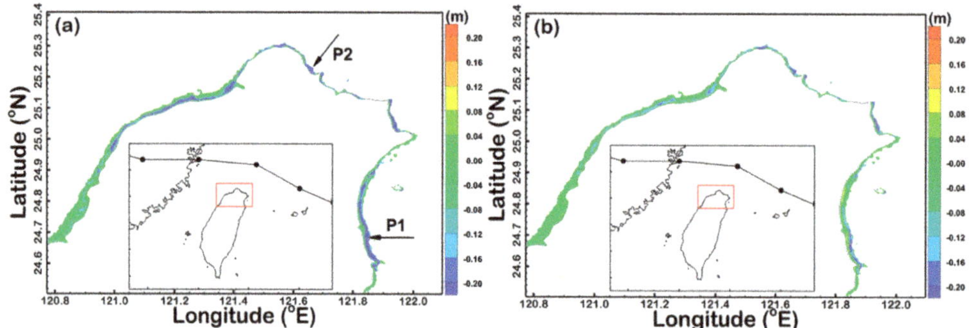

Figure 13. Spatial distribution of the difference in the maximum SWH between the scenarios of (**a**) S_NO1 and S_NO4 and (**b**) S_NO1 and S_NO5 during the passage of Super Typhoon Maria in 2018. The areas with water depths greater than -20 m are shown.

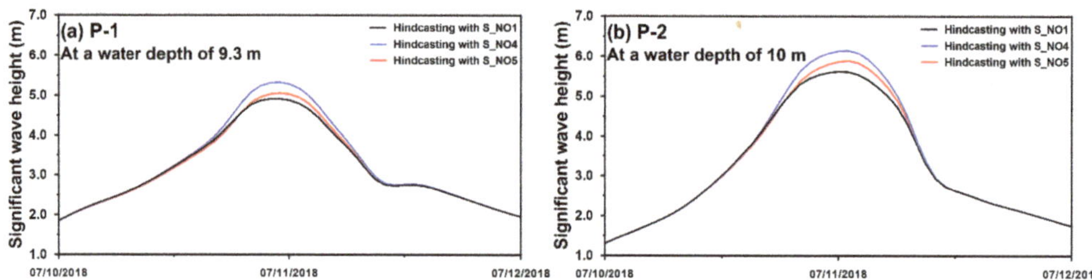

Figure 14. Time series of hindcasted SWHs for (**a**) P1 and (**b**) P2 using the S_NO1, S_NO4 and S_NO5 scenarios during the passage of Super Typhoon Maria in 2018.

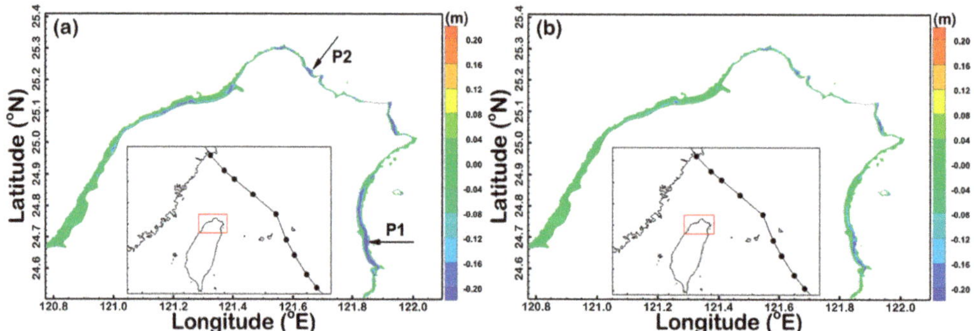

Figure 15. Spatial distribution of the difference in the maximum SWH between the scenarios of (**a**) S_NO1 and S_NO4 and (**b**) S_NO1 and S_NO5 during the passage of Super Typhoon Lekima in 2019. The areas with water depths greater than -20 m are shown.

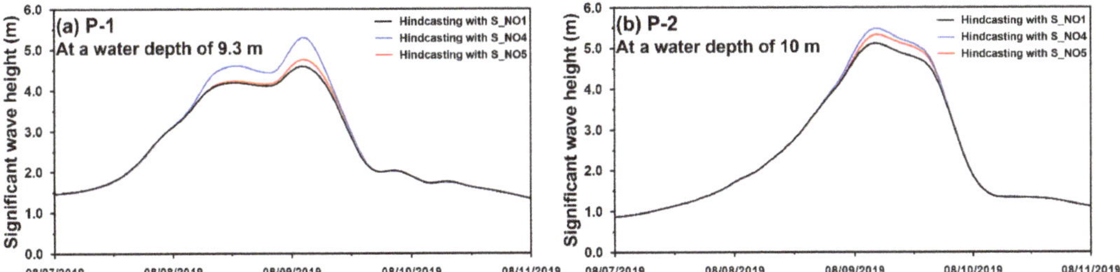

Figure 16. Time series of hindcasted SWHs for (**a**) P1 and (**b**) P2 using the S_NO1, S_NO4 and S_NO5 scenarios during the passage of Super Typhoon Lekima in 2019.

4. Discussion and Uncertainty

Hindcasting an accurate storm wave height is highly dependent on the accuracy of the typhoon wind fields. Reanalysis wind products underestimate typhoon winds, e.g., Çalışır et al. [53] evaluated the quality of ERA5 and CFSR (Climate Forecast System Reanalysis) winds and the contribution of reanalysis wind products to a wave modeling performance in a semi-closed sea. Their results revealed that ERA5 and CFSR tend to underestimate wind speeds, and ERA5 performs worse than CFSR at higher wind speeds (such as typhoon winds) and better at lower wind speeds. Thus, the utilization of hybrid winds through the superposition method (the combination of reanalysis wind products and parametric cyclone wind models) or the direct modification method (the combination of reanalysis wind products and the maximum wind speeds of typhoons from the best track data) is the most advantageous to consider storm wave hindcasting in both near-field and far-field regions of the typhoon's center. However, uncertainty remains to be clarified; that is, the radius of the modified scale (R_{trs}) cannot be formulated universally for each typhoon. Owing to the W_{Bmax} is higher than W_{Emax}, the area with stronger winds of a typhoon is extended as R_{trs} increases (according to Equation (1)). This expansion allows the hindcasted SWHs to grow earlier and attenuate later (i.e., larger) than the measurements (as shown in Figures 3 and 6). Additionally, as shown in Figure 3b,c, the measured peak SWH at the Suao wave buoy occurred two hours earlier than that at the Longdong wave buoy. However, the occurrence time of hindcasted peak SWH at the Suao wave buoy coincided with that at the Longdong wave buoy. This phenomenon might be due to the low spatial resolution of the original ERA5 winds (at roughly 31 km), and the periphery circulation of a typhoon cannot be resolved with such a coarse spatial resolution. Although the wind field derived from $R_{trs} = 4R_{max}$ is employed for designing a series of model experiments based on limited case studies, further studies are still needed to verify it.

Fully coupled ocean circulation and spectral wave numerical modeling systems are undergoing widespread use for all types of regional applications, such as operational predictions, wave climate evaluations, extreme storm waves and surge analyses. However, the wave-induced hydrodynamics during the period of typhoons in shallow nearshore waters simulated by these modeling systems remain uncertain. The scarcity of field observations for wave parameters in the surf zone to verify the modeling systems is one of the most important factors leading to uncertainty.

The SWH simulations in the surf zone are more sensitive to the various wave-breaking formulations than the various wave-breaking criteria. To reconfirm the result obtained from Section 3.2, the spatial distribution of the difference in the maximum SWH between the scenarios of S_NO1 (wave-breaking formulation of BJ87 with constant wave-breaking criteria) and S_NO2 (wave-breaking formulation of TG83 with constant wave-breaking criteria) and S_NO1 and S_NO3 (wave-breaking formulation of CT93 with constant wave-breaking criteria) in the surf zone (sea areas with water depths greater than −20 m are shown) of southeastern China for Super Typhoons Maria in 2018 and Lekima in 2019 are shown in Figures 17 and 18, respectively. As seen in Figures 17 and 18, the differences in maximum SWH between the S_NO1 and S_NO3 scenarios (Figures 17b and 18b) are higher than those between the S_NO1 and S_NO2 scenarios (Figures 17a and 18a) for both typhoons. Additionally, the surf zones with significant differences in maximum SWH resulting from the various wave-breaking formulations occur on the right side of the typhoons where the wind speeds are stronger. These findings are identical to the result derived from Section 3.2 and the reports from previous studies [54–57].

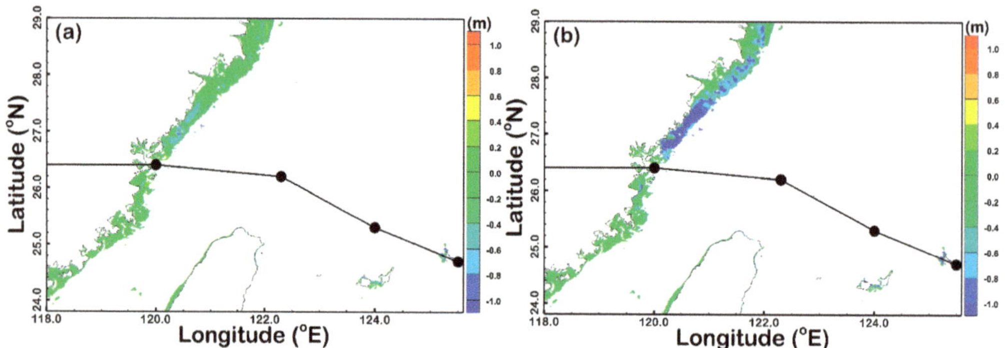

Figure 17. Spatial distribution of the difference in maximum SWH between the scenarios of (**a**) S_NO1 and S_NO2 and (**b**) S_NO1 and S_NO3 when Super Typhoon Maria (2018) made landfall on the southeast coast of China. The areas with water depths greater than −20 m are shown.

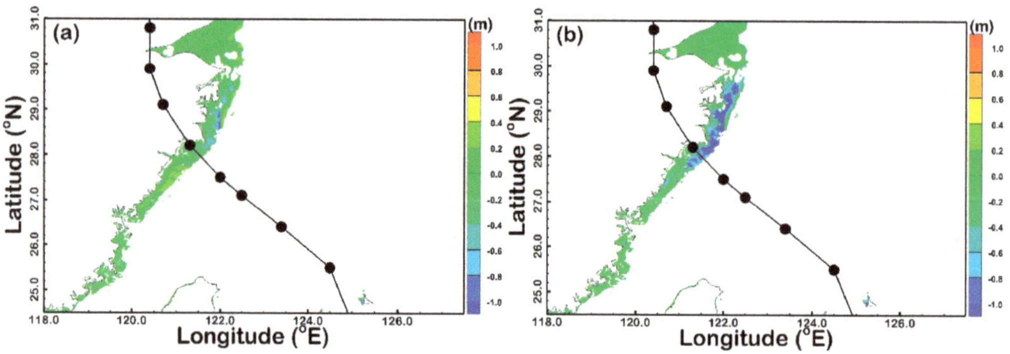

Figure 18. Spatial distribution of the difference in maximum SWH between the scenarios of (**a**) S_NO1 and S_NO2 and (**b**) S_NO1 and S_NO3 when Super Typhoon Lekima (2019) made landfall on the southeast coast of China. The areas with water depths greater than −20 m are shown.

5. Summary and Conclusions

In this paper, the effects of wave breaking formulations and wave breaking criteria in hindcasting typhoon-driven storm waves are investigated for shallow nearshore waters off northern Taiwan. A fully coupled high-resolution, unstructured grid wave-circulation modeling system, SCHISM-WWM-III, with a large computational domain was applied to hindcast the wind waves caused by the passages of Super Typhoon Maria in 2018 and Super Typhoon Lekima in 2019. The ERA5 reanalysis product were merged with the maximum wind from the best track dataset, and the hybrid typhoon wind created and served as the meteorological conditions for the SCHISM-WWM-III modeling system through a direct modification technique. The hindcasted SWHs during the typhoon period were more sensitive to the radius of the modified scale, R_{trs} and this phenomenon was more pronounced at the peak SWH. R_{trs} equal to four times R_{max} (the radius at the maximum typhoon wind speed) is an adequate radius of the modified scale for simulating the storm

J. Mar. Sci. Eng. **2021**, *9*, 706

waves induced by Super Typhoon Maria in 2018 and Super Typhoon Lekima in 2019 via model validations. A series of numerical experiments were conducted using the SCHISM-WWM-III modeling system incorporated with modified typhoon winds to better understand the wave hydrodynamics in the shallow nearshore waters off northern Taiwan. The results derived from the designed numerical experiments reveal that the wave breaking formulations influence the hindcast of storm waves in the surf zone of northern Taiwan. The maximum difference in peak SWH could reach 2.5 m and 1.2 m for Super Typhoons Maria (2018) and Lekima (2019), respectively, when the wave-breaking formulations of BJ78 and CT93 were introduced to the SCHISM-WWM-III modeling system. Regarding the wave-breaking criterion on the hindcast of typhoon waves in the surf zone of northern Taiwan, compared with the wave-breaking formulation, the maximum difference in peak SWHs was relatively non-sensitive to the wave-breaking criterion. The maximum difference in peak SWH is only 0.5 m using the constant breaking criterion, the breaking criterion based on local steepness or the breaking criterion based on peak steepness. Another important finding is that the utilization of the BJ78 wave-breaking formulation usually underpredicts the typhoon-generated SWHs in shallow nearshore waters than other parametrizations (i.e., TG83 and CT93). In future research, it will be important to acquire measurements in the surf zone where wave hydrodynamics are more sensitive to wave-breaking formulations and criteria during the passage or landfall of typhoons.

Author Contributions: Conceptualization, S.-C.H. and W.-B.C.; data curation, H.-L.W. and W.-D.G.; formal analysis, W.-D.G.; investigation, W.-B.C.; methodology, H.-L.W. and W.-B.C.; software, C.-H.C. and W.-R.S.; supervision, S.-C.H.; writing–original draft, W.-B.C.; Writing–review and editing, S.-C.H. All authors have read and agreed to the published version of the manuscript.

Funding: This research was supported by the Ministry of Science and Technology (MOST), Taiwan, grant No. MOST 109-2221-E-865-001.

Institutional Review Board Statement: Not applicable.

Informed Consent Statement: Not applicable.

Data Availability Statement: Publicly available datasets were analyzed in the present study. The data can be found here: https://ocean.cwb.gov.tw/V2/ (accessed date on 26 June 2021).

Acknowledgments: The authors thank the Central Weather Bureau, Ministry of Transportation and Communications, Taiwan, for providing the survey data and Joseph Zhang at the Virginia Institute of Marine Science, College of William and Mary, for kindly sharing his experiences using the numerical model.

Conflicts of Interest: The authors declare no conflict of interest.

References

1. Reguero, B.G.; Losada, I.J.; Méndez, F.J. A recent increase in global wave power as a consequence of oceanic warming. *Nat. Commun.* **2019**, *10*, 205. [CrossRef]
2. Stockdon, H.F.; Sallenger, A.H.; Holman, R.A.; Howd, P.A. A simple model for the spatially-variable coastal response to hurricanes. *Mar. Geol.* **2007**, *238*, 1–20. [CrossRef]
3. Hoeke, R.K.; McInnes, K.L.; Kruger, J.C.; McNaught, R.J.; Hunter, J.R.; Smithers, S.G. Widespread inundation of Pacific islands triggered by distant-source wind-waves. *Glob. Planet. Chang.* **2013**, *108*, 128–138. [CrossRef]
4. Wadey, M.; Brown, S.; Nicholls, R.J.; Haigh, I. Coastal flooding in the Maldives: An assessment of historic events and their implications. *Nat. Hazards* **2017**, *89*, 131–159. [CrossRef]
5. Melet, A.; Almar, R.; Hemer, M.; le Cozannet, G.; Meyssignac, B.; Ruggiero, P. Contribution of wave setup to projected coastal sea level changes. *J. Geophys. Res. Ocean.* **2020**, *125*, e2020JC016078. [CrossRef]
6. Hoefel, F.; Elgar, S. Wave-induced sediment transport and sandbar migration. *Science* **2003**, *299*, 1885–1887. [CrossRef] [PubMed]
7. Feddersen, F. Scaling surf zone turbulence. *Geophys. Res. Lett.* **2012**, *39*, L18613. [CrossRef]
8. Longuet-Higgins, M.S.; Stewart, R.W. Radiation stresses in water waves; a physical discussion, with applications. *Deep Sea Res.* **1964**, *11*, 529–562. [CrossRef]
9. Battjes, J.A.; Stive, M.J.F. Calibration and verification of a dissipation model for random breaking waves. *J. Geophys. Res.* **1985**, *90*, 9159–9167. [CrossRef]

10. Nairn, R.B. Prediction of Cross-Shore Sediment Transport and Beach Profile Evolution. Ph.D. Thesis, Imperial College London, London, UK, 1990.

11. Holthuijsen, L.H.; Booij, N. Experimental Wave Breaking in SWAN. In *Coastal Engineering 2006*; McKee Smith, J., Ed.; American Society of Civil Engineers: New York, NY, USA, 2006; pp. 392–402.

12. Apotsos, A.; Raubenheimer, B.; Elgar, S.; Guza, R.T. Testing and calibrating parametric wave transformation models on natural beaches. *Coast. Eng.* **2008**, *55*, 224–235. [CrossRef]

13. Ruessink, B.G.; Walstra, D.J.R.; Southgate, H.N. Calibration and verification of a parametric wave model on barred beaches. *Coast. Eng.* **2003**, *48*, 139–149. [CrossRef]

14. Jelesnianski, C.P. A numerical calculation of storm tides induced by a tropical storm impinging on a continental shelf. *Mon. Weather Rev.* **1965**, *93*, 343–358. [CrossRef]

15. Jelesnianski, C.P. Numerical computations of storm surges without bottom stress. *Mon. Weather Rev.* **1996**, *94*, 379–394. [CrossRef]

16. Dube, S.K.; Sinha, P.C.; Roy, G.D. The numerical simulation of storm surges along the Bangladesh coast. *Dyn. Atmos. Ocean.* **1985**, *9*, 121–133. [CrossRef]

17. Ginis, I.; Sutyrin, G. Hurricane-generated depth-averaged currents and sea surface elevation. *J. Phys. Oceanogr.* **1995**, *25*, 1218–1242. [CrossRef]

18. Lee, T.L. Back-propagation neural network for the prediction of the short-term storm surge in Taichung harbor, Taiwan. *Eng. Appl. Artif. Intell.* **2008**, *21*, 63–72. [CrossRef]

19. Chen, W.-B.; Liu, W.-C.; Hsu, M.H. Computational investigation of typhoon-induced storm surges along the coast of Taiwan. *Nat. Hazards* **2012**, *64*, 1161–1185. [CrossRef]

20. Pan, Y.; Chen, Y.P.; Li, J.X.; Ding, X.L. Improvement of wind field hindcasts for tropical cyclones. *Water Sci. Eng.* **2016**, *9*, 58–66. [CrossRef]

21. Chen, W.-B.; Chen, H.; Hsiao, S.-C.; Chang, C.-H.; Lin, L.-Y. Wind forcing effect on hindcasting of typhoon-driven extreme waves. *Ocean Eng.* **2019**, *188*, 106260. [CrossRef]

22. Hsiao, S.-C.; Chen, H.; Chen, W.-B.; Chang, C.-H.; Lin, L.-Y. Quantifying the contribution of nonlinear interactions to storm tide simulations during a super typhoon event. *Ocean Eng.* **2019**, *194*, 106661. [CrossRef]

23. Hsiao, S.-C.; Chen, H.; Wu, H.-L.; Chen, W.-B.; Chang, C.-H.; Guo, W.-D.; Chen, Y.-M.; Lin, L.-Y. Numerical Simulation of Large Wave Heights from Super Typhoon Nepartak (2016) in the Eastern Waters of Taiwan. *J. Mar. Sci. Eng.* **2020**, *8*, 217. [CrossRef]

24. Hsiao, S.-C.; Wu, H.-L.; Chen, W.-B.; Chang, C.-H.; Lin, L.-Y. On the Sensitivity of Typhoon Wave Simulations to Tidal Elevation and Current. *J. Mar. Sci. Eng.* **2020**, *8*, 731. [CrossRef]

25. Knaff, J.A.; Sampson, C.R.; Demaria, M.; Marchok, T.P.; Gross, J.M.; Mcadie, C.J. Statistical Tropical Cyclone Wind Radii Prediction Using Climatology and Persistence. *Weather Forecast.* **2007**, *22*, 781–791. [CrossRef]

26. Zhang, Y.J.; Ye, F.; Stanev, E.V.; Grashorn, S. Seamless cross-scale modelling with SCHISM. *Ocean Modell.* **2016**, *102*, 64–81. [CrossRef]

27. Zhang, Y.J.; Baptista, A.M. SELFE: A semi-implicit Eulerian-Lagrangian finite-element model for cross-scale ocean circulation. *Ocean Modell.* **2008**, *21*, 71–96. [CrossRef]

28. Zhang, Y.J.; Ye, F.; Yu, H.; Sun, W.; Moghimi, S.; Myers, E.; Nunez, K.; Zhang, R.; Wang, H.; Roland, A.; et al. Simulating compound flooding events in a hurricane. *Ocean Dyn.* **2020**, *70*, 621–640. [CrossRef]

29. Shchepetkin, A.F.; McWilliams, J.C. The regional oceanic modeling system (ROMS): A split-explicit, free-surface, topography-following-coordinate, oceanic model. *Ocean Model.* **2005**, *9*, 347–404. [CrossRef]

30. Chen, W.-B.; Liu, W.-C. Modeling flood inundation induced by river flow and storm surges over a river basin. *Water* **2014**, *6*, 3182–3199. [CrossRef]

31. Chen, W.-B.; Liu, W.-C. Assessment of storm surge inundation and potential hazard maps for the southern coast of Taiwan. *Nat. Hazards* **2016**, *82*, 591–616. [CrossRef]

32. Chen, W.-B.; Liu, W.-C.; Hsu, M.-H. Modeling evaluation of tidal stream energy and the impacts of energy extraction on hydrodynamics in the Taiwan strait. *Energies* **2013**, *6*, 2191–2203. [CrossRef]

33. Chen, W.-B.; Chen, H.; Lin, L.-Y.; Yu, Y.-C. Tidal Current Power Resource and Influence of Sea-Level Rise in the Coastal Waters of Kinmen Island, Taiwan. *Energies* **2017**, *10*, 652. [CrossRef]

34. Chen, W.-B.; Liu, W.-C.; Hsu, M.-H.; Hwang, C.-C. Modeling investigation of suspended sediment transport in a tidal estuary using a three-dimensional model. *Appl. Math. Model.* **2015**, *39*, 2570–2586. [CrossRef]

35. Chen, W.-B.; Liu, W.-C. Investigating the fate and transport of fecal coliform contamination in a tidal estuarine system using a three-dimensional model. *Mar. Pollut. Bull.* **2017**, *116*, 365–384. [CrossRef]

36. Roland, A. Development of WWM II: Spectral Wave Modeling on Unstructured Meshes. Ph.D. Thesis, Technology University Darmstadt, Darmstadt, Germany, 2009.

37. Komen, G.J.; Cavaleri, M.; Donelan, K.; Hasselmann, S.; Hasselmann, P.A.E.; Janssen, M. *Dynamics and Modelling of Ocean Waves*; Cambridge University Press: Cambridge, UK, 1994; p. 532.

38. Hasselmann, K.; Barnett, T.P.; Bouws, E.; Carlson, H.; Cartwright, D.E.; Enke, K.; Ewing, J.A.; Gienapp, H.; Hasselmann, D.E.; Kruseman, P.; et al. *Measurements of Wind-Wave Growth and Swell Decay during the Joint North Sea Wave Project (JONSWAP)*; Deutsches Hydrographisches Institut: Berlin, Germany; Hamburg, Germany, 1973.

39. Chen, W.-B.; Lin, L.-Y.; Jang, J.-H.; Chang, C.-H. Simulation of typhoon-induced storm tides and wind waves for the northeastern coast of Taiwan using a tide-surge-wave coupled model. *Water* **2017**, *9*, 549. [CrossRef]
40. Su, W.-R.; Chen, H.; Chen, W.-B.; Chang, C.-H.; Lin, L.-Y.; Jang, J.-H.; Yu, Y.-C. Numerical investigation of wave energy resources and hotspots in the surrounding waters of Taiwan. *Renew. Energy* **2018**, *118*, 814–824. [CrossRef]
41. Shih, H.-J.; Chang, C.-H.; Chen, W.-B.; Lin, L.-Y. Identifying the Optimal Offshore Areas for Wave Energy Converter Deployments in Taiwanese Waters Based on 12-Year Model Hindcasts. *Energies* **2018**, *11*, 499. [CrossRef]
42. Hsiao, S.-C.; Cheng, C.-T.; Chang, T.-Y.; Chen, W.B.; Wu, H.-L.; Jang, J.-H.; Lin, L.-Y. Assessment of Offshore Wave Energy Resources in Taiwan Using Long-Term Dynamically Downscaled Winds from a Third-Generation Reanalysis Product. *Energies* **2021**, *14*, 653. [CrossRef]
43. Orton, P.; Georgas, N.; Blumberg, A.; Pullen, J. Detailed modeling of recent severe storm tides in estuaries of the New York City region. *J. Geophys. Res. Ocean.* **2012**, *117*, C09030. [CrossRef]
44. Zheng, L.; Weisberg, R.H.; Huang, Y.; Luettich, R.A.; Westerink, J.J.; Kerr, P.C.; Donahue, A.S.; Grane, G.; Akli, L. Implications from the comparisons between two- and three-dimensional model simulations of the Hurricane Ike storm surge. *J. Geophys. Res. Ocean.* **2013**, *118*, 3350–3369. [CrossRef]
45. Zu, T.; Gana, J.; Erofeevac, S.Y. Numerical study of the tide and tidal dynamics in the South China Sea. *Deep Sea Res. Part I* **2008**, *55*, 137–154. [CrossRef]
46. Liu, Z.; Wang, H.; Zhang, Y.J.; Magnusson, L.; Loftis, J.D. Cross-scale Modeling of Storm Surge, Tide and inundation in Mid-Atlantic Bight and New York City during Hurricane Sandy, 2012. *Estuar. Coast. Shelf Sci.* **2020**, *233*, 106544. [CrossRef]
47. Battjes, J.A.; Beji, S. Breaking Waves Propagating over a Shoal. In *Coastal Engineering 1992, Proceedings of the 23rd International Conference on Coastal Engineering, Venice, Italy, 4–9 October 1992*; ASCE: Reston, VA, USA, 1992; pp. 42–50.
48. Battjes, J.A.; Janssen, J.P.F.M. Energy Loss and Set-Up Due to Breaking of Random Waves. In *Coastal Engineering 1978, Proceedings of the 16th International Conference on Coastal Engineering, Hamburg, Germany, 27 August–3 September 1978*; ASCE: Reston, VA, USA, 1978; pp. 569–587.
49. Thornton, E.B.; Guza, R.T. Transformation of wave height distribution. *J. Geophys. Res.* **1983**, *88*, 5925–5938. [CrossRef]
50. Thornton, E.B.; Guza, R.T. Energy saturation and phase speeds measured on a natural beach. *J. Geophys. Res.* **1982**, *87*, 9499–9508. [CrossRef]
51. Church, J.C.; Thornton, E.B. Effects of breaking wave induced turbulence within a longshore current model. *Coast. Eng.* **1993**, *20*, 1–28. [CrossRef]
52. Miche, R. Mouvements ondulatoires des mers en profondeur constante ou décroissante. *Annales des Ponts Chaussees* **1944**, 369–406.
53. Calisir, E.; Soran, M.B.; Akpinar, A. Quality of the ERA5 and CFSR winds and their contribution to wave modelling performance in a semi-closed sea. *J. Oper. Oceanogr.* **2021**. [CrossRef]
54. Van der Westhuysen, A.J. Modeling of depth-induced wave breaking under finite depth wave growth conditions. *J. Geophys. Res. Ocean* **2009**, *115*, C01008. [CrossRef]
55. Salmon, J.E.; Holthuijsen, L.H. Modeling depth-induced wave breaking over complex coastal bathymetries. *Coast. Eng.* **2015**, *105*, 21–35. [CrossRef]
56. Lin, S.; Sheng, J. Assessing the performance of wave breaking parameterizations in shallow waters in spectral wave models. *Ocean Modell.* **2017**, *120*, 41–59. [CrossRef]
57. Pezerat, M.; Bertin, X.; Martins, K.; Mengual, B.; Hamm, L. Simulating storm waves in the nearshore area using spectral model: Current issues and a pragmatic solution. *Ocean Modell.* **2021**, *158*, 101737. [CrossRef]

Journal of
Marine Science and Engineering

Article

The Focusing Waves Induced by Bragg Resonance with V-Shaped Undulating Bottom

Haiming Zhang [1,2], Aifeng Tao [1,2,*], Junhao Tu [1,2], Junwei Su [1,2] and Shuya Xie [1,2]

[1] Key Laboratory of Ministry of Education for Coastal Disaster and Protection, Hohai University, Nanjing 210024, China; haiming@hhu.edu.cn (H.Z.); tu54@hhu.edu.cn (J.T.); 171303020052@hhu.edu.cn (J.S.); Xieshuya@hhu.edu.cn (S.X.)

[2] College of Harbour, Coastal and Offshore Engineering, Hohai University, Nanjing 210024, China

* Correspondence: aftao@hhu.edu.cn; Tel.: +86-15295540618

Abstract: Intensive wave reflection occurs when the wavelengths of the incident waves and bottom undulations are in a 2:1 ratio. Existing studies have included the Bragg resonance phenomenon of waves passing over a continuous undulating bottom parallel to and oblique to the shoreline. More generally, the Bragg resonance mechanism is used as a means of coastal protection, rather than wave power generation. To focus the wave energy in a specific area, here, we propose sinusoidal sandbars of a horizontal V-shaped pattern, which is formed by two continuous undulating bottoms inclined at an angle to each other and the center axis of the angle is perpendicular to the shoreline. Based on the high-order spectral (HOS) numerical model, both the characteristics of Bragg resonance induced by the regular waves and random waves are investigated. In the scenario of regular waves, it shows that the wave-focusing effect is related to the angle of the V-shaped undulating bottom, and the optimal angle of inclination for the V-shaped undulating bottom is 162.24°. On that basis, considering the interactions between the random waves and the V-shaped undulating bottom of 162.24°, the Bragg resonance characteristics of random waves are studied. The *BFI* factor combining wave steepness and spectrum width can evaluate the focusing intensity of the Bragg resonance of the random waves. For *BFI*, in the range of 0.15–1.0, the values of H_{smax}/H_{s0} linearly increase with the increase of *BFI*.

Keywords: three-dimensional Bragg resonance; regular waves; random waves; high-order spectral (HOS) method; Gaussian spectrum; V-shaped undulating bottom

Citation: Zhang, H.; Tao, A.; Tu, J.; Su, J.; Xie, S. The Focusing Waves Induced by Bragg Resonance with V-Shaped Undulating Bottom. *J. Mar. Sci. Eng.* **2021**, *9*, 708. https:// doi.org/10.3390/jmse9070708

Academic Editors: Shih-Chun Hsiao, Wen-Son Chiang and Wei-Bo Chen

Received: 19 May 2021
Accepted: 24 June 2021
Published: 27 June 2021

Publisher's Note: MDPI stays neutral with regard to jurisdictional claims in published maps and institutional affiliations.

1. Introduction

Ocean waves contain large untapped renewable energy sources, which could be used to fulfil the energy demand [1]. Wave energy has the characteristics of high power density and wide distribution, and is renewable [2]. Hitherto, about one thousand wave energy converter (WEC) inventions have been patented [3], and more than 200 of them have entered the model testing stage [2]. However, present research mainly focuses on promoting the conversion efficiency of wave energy converters by improved mechanical design [4]. There are few research results to improve the wave energy generation efficiency from the perspective of increasing the wave energy density of the target sea area. Taking the characteristics of Bragg resonance into account is a good idea for wave energy focusing.

Bragg reflection or resonance is initially referred to as a special physical optical phenomenon of X-rays [5]. In the 1980s, Davies [6] first studied that Bragg resonance occurs if the bottom wavelength is an integral multiple of half wavelength of the incident waves. A followed idea originated for coastal protection was put forward. The strong reflection of water waves by seabed bars was investigated experimentally [7,8] as well as theoretically [9–11]. Afterwards, Bailard et al. [12] have extensively studied theoretically, computationally and experimentally to learn explicitly the feasibility of using longshore seabed bars to protect the shore and summarized its effectiveness and limitations. Some issues, including the partially standing-wave pattern [13] and shoreward increase of wave amplitude [14–16] by Bragg reflection, were exposed. Howard & Yu [17] and Weidman et al. [18] have demonstrated that a specific phase relationship of wave-bottom interactions is significant, and sometimes it can cause exponentially varying standing-wave patterns. It is illustrated that Bragg resonance leads to offward intense reflection, even shoreward wave amplitude increase. Considering the strong effect of Bragg resonance, more studies focus on coastal protection or beach erosion from the aspect of the backward of the seabed bars.

Subsequent investigations of Bragg resonance mechanism in the field of coastal protection and engineering application pay more attention to the actual situations, e.g., oblique incident waves [10,19–22], sub- and superharmonic frequency [19–23], and seabed configuration [10,21,22,24].

Recently, Couston et al. [22] have proposed to revamp the Bragg resonance mechanism as a means of coastal protection by considering oblique seabed bars that divert, rather than reflect, shore-normal incident waves to the shore-parallel direction. Indeed, the novelty that with two superposed sets of oblique seabed bars reflecting waves along the shore-parallel direction, becomes efficiently deflected far to the sides, leaving a wake of decreased wave activity downstream of the patch. However, the wave characteristics of the forward of the oblique seabed bars are not studied by Couston et al. By contrast, few studies concern the wave reflection of the forward. From another perspective of the seabed bars reflecting waves, the wave-bottom interactions due to Bragg resonance may be of interest and be more in accordance with wave power generation need.

In the cases where closed-form solutions cannot be obtained, a wide range of numerical models are available to study water-wave scattering by seabed topographies [22]. These include the extended versions, e.g., the mild-slope equations [11,25], the coupled-mode approach [26,27], the integral matching/discretized bottom method [28,29], the fully nonlinear Boussinesq equation adjusted for rapid bottom undulations [30], an asymptotic linear analytical solution (ALAS) [24], and the high-order spectral method [31]. Numerical investigations of Bragg scattering have helped explain several discrepancies between theory and experiments, including the difference between the observed and predicted Class II Bragg resonance frequency due to evanescent modes [32] and the resonant frequency downshift/upshift for the subharmonic/superharmonic Class III Bragg condition due to high-order nonlinearity [30].

Simultaneously, Bragg resonance theory has made some achievements in application fields. A series of submerged breakwaters have been developed to achieve coastal protection based on the Bragg resonance characteristic of reflecting waves to the sea in recent years. For the two-dimensional numerical cases, the parameters' influence on the Bragg resonance reflection coefficient has been investigated by scholars from different aspects, e.g., the shape, height, width, the number of the submerged breakwaters, the slope of the seabed, and wave conditions [33–37]. Generally speaking, these numerical simulations are based on various theoretical methods to study normally incident waves over a series of rectangular or trapezoidal breakwaters. Ning et al. [38] investigated numerically and experimentally the effect of the breakwater shape on hydrodynamic behavior in a two-dimensional flume. Shih & Weng [39] studied the interactions between waves and various combinations of undulating breakwaters in a three-dimensional basin. The reflection coefficient, transmission coefficient, and attenuation of wave energy were analyzed. Liu et al. [40] examined the Bragg reflection of waves by multiple submerged semi-circular breakwaters, by considering obliquely and normally incident waves independently. Although two-dimensional flume or three-dimensional basin experimental studies have been carried out, the study based on Bragg resonance for improving wave energy generation has not been widely considered. At present, research on the use of wave energy is gradually carried out. Water flow and wave propagation can be altered by artificial terrains or control devices. Zheng et al. [41] carried out a laboratory study on wave-induced setup and wave-driven current in a 2DH reef-lagoon-channel system. Elandt et al. [42] found that gravity waves can focus at a specific location by a concave mirror or a convex lens of gravity waves. Tao et al. [43] investigated experimentally strong wave reflection induced by Bragg resonance. They found that wave reflection is effective to amplify the free-surface oscillation amplitude and focus the wave energy in front of the undulating bottom. Zhang & Ning [44] confirmed that reflected waves from the parabolic opening of the breakwater can travel towards a fixed focus position. For a specified wave environment, the wave heights at the focus positions can reach over several times of the incident wave heights, indicating wave energy multiplication. They all proposed the concept of the focal point. Instead of putting a multitude of wave energy harvesting devices over a large area, a large wave energy absorber can be placed at the focal point. However, the influence of peculiar spatial bottom layouts is ignored, which also affect the focal point and focusing area. Previous physical experiment research [43] shows that waves are strongly reflected in front of the undulating bottom. Based on that, this study considers the wave-focusing effect in a specific area by optimizing the undulating bottom pattern.

The objective of this study is to investigate both the characteristics of Bragg resonance induced by the regular waves and random waves. Three main research questions need to be answered: (1) How does the angle of inclination for the V-shaped undulating bottom affect the wave-focusing characteristics induced by the regular waves? (2) What analytical methods are used to determine the optimal angle of the V-shaped undulating bottom? (3) How do the steepness and spectrum width of random waves affect the wave-focusing characteristics with the V-shaped undulating bottom, and what is the law of their influence on wave height? By the high-order spectral (HOS) numerical simulation, these questions have been examined. The concept of V-shaped undulating bottom is first proposed, i.e., a horizontal V-shaped pattern with two continuous undulating bottoms inclined at the same angle perpendicular to the shoreline. The V-shaped undulating bottom can not only play the role of reflecting waves in front of the undulations, but also exploit the advantage of symmetrical "V-shaped", which can focus waves on the central axis and increase the wave heights in the focusing areas. However, waves in the ocean are stochastic, and the random waves contain wave components of different frequencies, each of which has different Bragg resonance effect due to interactions with the undulating bottom of fixed frequency. Most of the previous studies on wave Bragg resonance only consider the regular waves at the dominant resonance frequency. In addition, there are few studies on Bragg resonance involving the interactions between random waves and undulating bottom. In this work, we have further studied the Bragg resonance characteristics of the interactions between random waves and V-shaped undulating bottom by the numerical simulation of HOS. This study has been organized as follows. In Section 2, the generalized Bragg conditions, the high-order spectral (HOS) method, and the model establishment are introduced. In Section 3, Bragg resonance reflection coefficients of regular waves are studied from the perspectives of different angles θ and different values f. In Section 4, the Bragg resonance focusing characteristics of regular waves and random waves for V-shaped undulating bottom are expounded. In Section 5, the conclusions are provided.

2. Methods

According to the high-order spectral (HOS) method [20], an effective computational method for general nonlinear wave-bottom (resonant) interactions and the boundary-value problem are stated.

2.1. Generalized Bragg Conditions

The mechanism for Bragg resonances is analogous to that for nonlinear (surface) wave-wave resonant interactions in the absence of bottom undulations. Thus, general Bragg conditions can be deduced from the well-known resonance condition for nonlinear wave-wave interactions [45]. For a wave field over uniform depth h, interactions among different wave components become resonant at order m (in wave steepness) if the wave numbers k_j and the corresponding frequencies ω_j satisfy:

$$\left.\begin{array}{l} k_1 \pm k_2 \pm \cdots \pm k_{m+1} = 0 \\ \omega_1 \pm \omega_2 \pm \cdots \pm \omega_{m+1} = 0 \end{array}\right\} (m \geq 2) \tag{1}$$

where the same combination of signs is to be taken in both equations, and k_j and ω_j satisfy the linear dispersion relation:

$$\omega_j^2 = g|k_j|\tanh|k_j|h \tag{2}$$

Generalized Bragg resonance conditions in the presence of bottom ripples are obtained by replacing one or more of the free-surface wave components in Equation (1) by periodic bottom ripple components of corresponding wavenumbers k_{bj} but with zero frequencies (since the ripples are fixed) [20]. Thus, by combining wavenumbers and frequencies of surface waves and bottom ripples, we obtain general conditions for Bragg resonances at each order, $m = 2, 3 \ldots$

Here, the Bragg resonance between free wave surface and V-shaped undulating bottom originated satisfies generalized Bragg conditions. In addition, we propose that the first case considers two surface wave components of an incident wavenumber k_1 and a reflected wavenumber k_2, propagating over a V-shaped undulating horizontal bottom containing a single wavenumber k_b (analogously to surface waves, which refers to a fixed sinusoidally varying bottom with crest lines normal to k_b). The first case belongs to Class I Bragg resonance. The second case considers random surface wave generated by the Gaussian spectrum, propagating over a V-shaped undulating horizontal bottom containing a single wavenumber k_b (analogously to surface waves, which refers to a fixed sinusoidally varying bottom with crest lines normal to k_b). In the paper, Class I Bragg resonance is considered, and numerical simulations all adopt the high-order spectral (HOS) method.

2.2. High-Order Spectral (HOS) Method

Combined with the advantages of Zakharov equations and mode-coupling ideas based on potential theory (with the application of the Fast Fourier Transform), the HOS method has been proved as a suitable, robust, and highly efficient numerical method for direct phase-resolved simulation of nonlinear ocean wave field evolution [46–48], wave nonlinear mechanism analysis [49–52] and nonlinear wave-wave interactions and wave-body interactions [31,53,54]. In this paper, the HOS numerical model for three-dimensional Bragg resonance is considered based on the HOS method for general nonlinear wave-bottom interactions developed by Liu [20]. In addition, the focusing properties of the three-dimensional wave field caused by Bragg resonance with waves passing through V-shaped undulating bottom are studied.

The difficulty of establishing a numerical model is to deal with free-surface boundary conditions. To simplify the calculation, the high-order spectral (HOS) method introduces the velocity potential of the free surface in reference to Zakharov theory [55], so the free-surface boundary condition can be written in the following form:

$$\left.\begin{array}{l} \eta_t + \nabla_x\eta \cdot \nabla_x\phi^s - (1 + \nabla_x\eta \cdot \nabla_x\eta)\phi_z(x,\eta,t) = 0 \\ \phi_t^s + g\eta + \frac{1}{2}\nabla_x\phi^s \cdot \nabla_x\phi^s - \frac{1}{2}(1 + \nabla_x\eta \cdot \nabla_x\eta)\phi_z^2(x,\eta,t) = 0 \end{array}\right\} \tag{3}$$

2.3. Model Establishment

The V-shaped undulating bottom is shown in Figure 1. It shows that the two continuous undulating bottoms of a horizontal V-shaped pattern, a set of oblique undulations k_{b1} (the upper part of the V-shaped pattern). To match with k_{b1}, the other set of oblique undulations k_{b2} (the lower part of the V-shaped pattern matched) is given. The values of k_{b1} and k_{b2} are the same for a given V-shaped undulating bottom, while the difference of the k_{b1} and k_{b2} is in the directions of the two vectors. The vectors of k_{b1} and k_{b2} are symmetric about the x-axis. The color vectors in Figure 1 describe simply the V-shaped undulating bottom configuration, where the number of the continuous sinusoidal undulations $N_b = 10$, an incident wave wavenumber k_1 (the direction of k_1 along the x-axis); two bottom undulations wavenumber k_{b1} and k_{b2} are perpendicular to crest lines of the each side of V-shaped undulating bottom respectively; the reflected waves k_{r1} and k_{r2} due to the two parts of bottom superpose each other ahead of the bottom. In addition, the relationship of the angle θ between k_{b1}/k_{b2} and x-axis and the angle α between the two sets of oblique crest lines satisfy:

$$\alpha = 180° - 2\theta \tag{4}$$

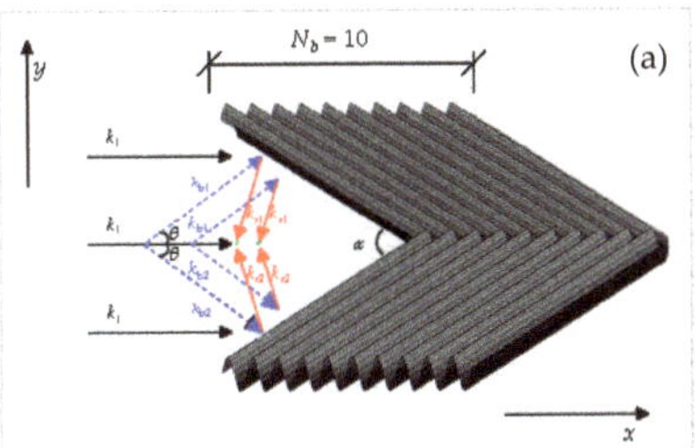
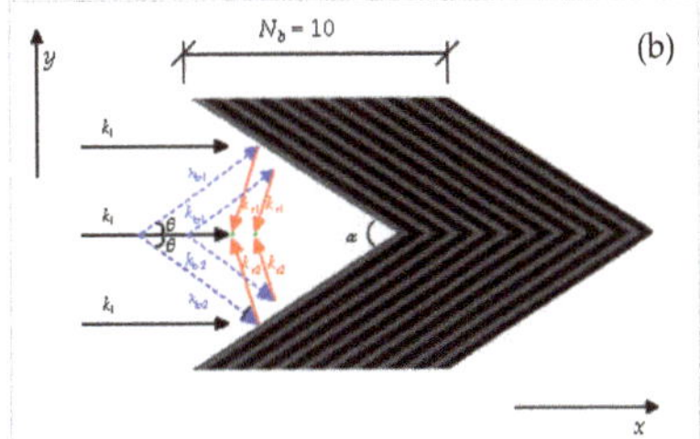

Figure 1. Schematic diagram of the relationship of the angle θ between k_{b1}/k_{b2} and x-axis and the angle α between the two sets of oblique crest lines: (**a**) Spatial graph of V-shaped undulating bottom ($\alpha < 180°$); (**b**) Plane graph of V-shaped undulating bottom ($\alpha < 180°$).

The following parameters for the HOS numerical model are given referring to reference [20]: the undulating bottom $k_b d = 0.31$, where d and k_b (k_b refers to k_{b1} and k_{b2}, and below is the same) are the amplitude and wavenumber of bottom undulations; the incident waves $k_1 A = 0.05$, where $2A \equiv \eta_{max} - \eta_{min}$, where A, k_1, η_{max} and η_{min} are the amplitude, wavenumber, maximum elevation, and minimum elevation for surface wave of the incident regular waves respectively; relative water depth $d/h = 0.16$, where d and h are the amplitude of bottom and water depth; the total numbers of bottom undulations $N_b = L_0/\lambda_b = 10$, where N_b, L_0, and λ_b are the number, total length, and wavelength of the sinusoidal undulating bottom respectively; the time step for the fourth-order Runge–Kutta integration Δt; the total simulation time T_s; running steps per period $T/\Delta t = 64$, where T is the period of the incident waves; the duration of the simulation periods $T_s/T = 20$; node numbers $N_x \times N_y = 512 \times 512$; nonlinear order $M = 3$; V-shaped undulating bottom angle $\alpha = 90°$–$180°$, where 24 values between 90° and 180° are distributed unevenly to the angle α of inclination for the V-shaped undulating bottom, as shown in Table 1.

Table 1. The simulation cases of V-shaped undulating bottom angles α.

No.	1	2	3	4	5	6	7	8	9	10	11	12
$\alpha(°)$	180.00	178.21	176.42	174.63	172.85	168.40	167.52	165.75	162.24	160.50	155.32	151.93
No.	13	14	15	16	17	18	19	20	21	22	23	24
$\alpha(°)$	148.58	145.29	140.47	137.33	132.74	126.87	119.93	115.99	112.21	106.26	98.65	90.00

3. Results

In this study, the three-dimensional Bragg resonance numerical model is validated from two aspects: the first one is to verify the regular wave reflection coefficients under different incident angles θ; the other one is to verify the regular wave reflection coefficients under different wavenumber ratios f (f is the ratio of 2 times the incident wavenumber ($2k_1$) to the undulating bottom wavenumber (k_b), i.e., $f = 2k_1/k_b$). This will prove the accuracy of the model in studying the focusing characteristics of Bragg resonance on V-shaped undulating bottom under different incident angles and different resonance wave numbers.

3.1. Bragg Resonance Reflection Coefficients of Regular Waves under Different Angles θ

The Bragg resonance reflection coefficient is the most important parameter to measure the magnitude of Bragg resonance. The Bragg resonance reflection coefficient is calculated according to the method proposed by Liu and Yue [20]. The Bragg reflection resonance coefficients $R(\theta)$ of the Class I Bragg resonance at different incident angles θ are simulated by the three-dimensional HOS method, as shown in Table 2, which are compared with the perturbation theory solutions of Mei [19] (oblique incidence on a finite strip of bars) in Figure 2.

Table 2. Reflection coefficients $R(\theta)$ of Class I Bragg resonance at different incident angles θ.

$\theta/(°)$	0	7.13	14.04	26.57	36.87	45.00	48.37	51.34
$R(\theta)$	0.72	0.70	0.68	0.54	0.29	0	0.09	0.25

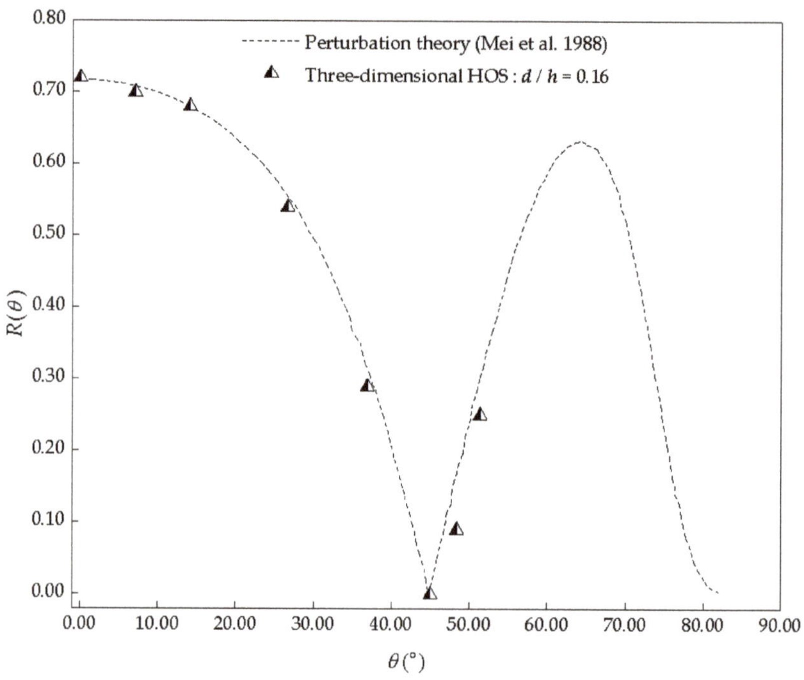

Figure 2. Reflection coefficients $R(\theta)$ comparisons of Class I Bragg resonance.

Figure 2 shows that the three-dimensional HOS method results are in good agreement with the perturbation theory solutions of Mei. When $\theta = 0°$ (normal incidence), $R(\theta)$ is the largest and up to 0.72. As θ increases, $R(\theta)$ first decreases with an increasing decaying rate. When θ reaches 45°, $R(\theta)$ decreases to 0, which means the wave propagation is not affected by the periodic undulating bottom. When $\theta > 45°$, $R(\theta)$ exhibits a second maximum point around $\theta = 64.5°$.

3.2. Bragg Resonance Reflection Coefficients of Regular Waves under Different Values f

The wavenumber ratios f is determined as 0.9–1.1 due to the study of Liu and Yue [20]. For f, in the range of 0.9–1.1, 17 different wavenumber ratios f are uniformly selected to investigate the regular wave reflection coefficients with different values of f under the forward ($\theta = 0°$) and oblique ($\theta = 19.5°$) incident conditions. The results of the three-dimensional HOS model and perturbation theory [19] are compared.

According to Figure 3, when $f = 0.9$–1.1, the Bragg resonance reflection coefficients increase first and then decrease with the increase of f. When f is slightly less than 1.0, the Bragg resonance reflection coefficients reach the peak, which also accords with the phenomenon of Bragg resonance frequency descending. The verification results confirm the reliability of the three-dimensional HOS numerical model.

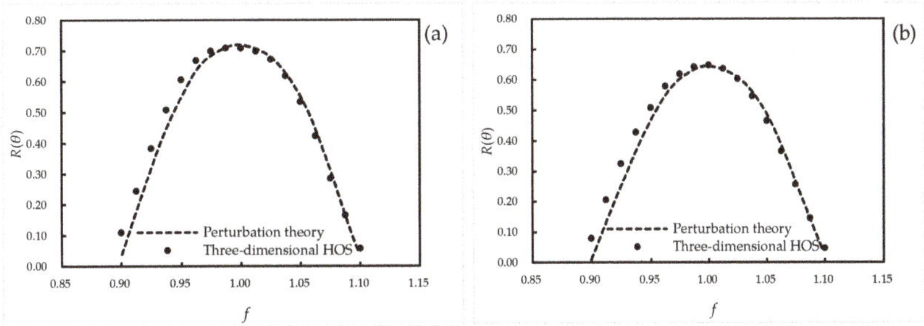

Figure 3. Bragg resonance reflection coefficients of regular waves under different wavenumber ratios f: (**a**) Forward incident condition ($\theta = 0°$); (**b**) Oblique incident condition ($\theta = 19.5°$).

4. Discussion

4.1. Bragg Resonance Focusing Characteristics of Regular Waves for V-Shaped Undulating Bottom

In the section, the wavelength of the incident wave is 1 m, which need 16 nodes to represent. The other parameters of the incident waves and undulating bottom can be calculated according to the section of the Model Establishment.

4.1.1. Spatial Distributions of Wave Amplitudes in the Focusing Areas for Regular Waves

The V-shaped layout can be regarded as the combination of two parts of oblique bottom undulations relative to the incident waves. The wave-focusing effects, therefore, are the effects of two parts of oblique bottom undulations relative to the incident waves. When the wave incident angle (between k_1 and k_b) θ is greater than $45°$, $R(\theta)$ has a secondary maximum. There is a relationship between α and θ, as shown in Figure 1.

As θ increases, the α of the V-shaped layout becomes smaller. As a result, the wave-focusing area becomes smaller, so the wave-focusing effect is weakened. The values of α parameter for the specific simulation cases are listed in Table 1.

Bragg resonance occurs when regular waves pass over the V-shaped undulating bottom and satisfy the conditions that the wavelengths of the incident waves and bottom undulations are in a 2:1 ratio. Incident waves and reflected waves from two directions superimposed in front of the bottom causes wave energy focusing effect. Under the condition of regular wave incidence with the same wave amplitude, the Bragg resonance effects of wave-bottom interactions at different V-shaped undulating bottom angles α are simulated by the three-dimensional HOS numerical model. The maximum wave amplitudes calculated (A_{max}) are compared with the incident wave amplitude (A_0), and the spatial distribution characteristics of wave amplitudes are analyzed by A_{max}/A_0. The maximum wave amplitudes calculated (A_{max}) are different at different node positions. So, the values of A_{max}/A_0 is also different. Table 3 shows that the calculated wave amplitude increases change at the focal points corresponding to different α ($90° \leq \alpha \leq 180°$). The node position of the maximum value of A_{max}/A_0 for each simulation case is called the focal point, and then A_{maxp}/A_0 is used to denote the value of A_{max}/A_0 at the focal point. When $\alpha = 180°$ (the incident wave direction k_1 is parallel to the bottom k_b), the calculated wave amplitude at the focal point increases to 1.81 times the initial incident wave amplitude. As θ increases, the α of the V-shaped layout becomes smaller. As a result, the wave-focusing area becomes smaller, so the wave-focusing effect is weakened. In conclusion, $90° \leq \alpha \leq 180°$ is mainly considered in this study.

J. Mar. Sci. Eng. **2021**, _9_, 708

Table 3. The calculated wave amplitude increases at the focal points corresponding to different angles α.

$\alpha(°)$	180.00	178.21	176.42	174.63	172.85	168.40	167.52	165.75
A_{maxp}/A_0	1.81	1.91	2.08	2.22	2.33	2.79	2.83	2.86
$\alpha(°)$	162.24	160.50	155.32	151.93	148.58	145.29	140.47	137.33
A_{maxp}/A_0	2.87	2.78	2.44	2.21	1.92	1.73	1.56	1.47
$\alpha(°)$	132.74	126.87	119.93	115.99	112.21	106.26	98.65	90.00
A_{maxp}/A_0	1.46	1.37	1.30	1.22	1.20	1.16	1.14	1.11

To better analyze the spatial distribution characteristics of wave amplitudes, $\alpha = 178.21°$, $168.40°$, $162.24°$, $155.32°$, $148.58°$, and $140.47°$ are selected for comparison. The spatial distributions of wave amplitudes in the focusing areas due to Bragg resonance at different V-shaped undulating bottom angles α are analyzed, the values of A_{max}/A_0 are shown in Figure 4.

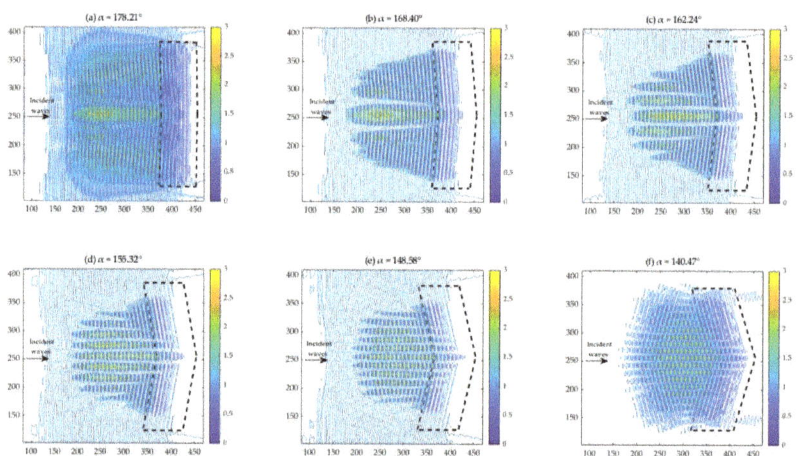

Figure 4. The values of A_{max}/A_0 in the scenarios of regular waves under different angles α of V-shaped undulating bottom (The dashed outline refers to the V-shaped undulating bottom range).

Figure 4 shows that no matter what degree α is, there are obvious wave-focusing areas and a series of focal points in front of the V-shaped undulating bottom. According to the principle of coastal protection, the wave amplitudes behind the bottom are possibly weakened owning to enormous waves reflected in front of the bottom. The V-shaped undulating bottom is generally symmetrical about the central axis; the wave-focusing areas are also symmetrical about the central axis. In addition, the wave-focusing areas are generally expand from right to left in front of the bottom in a "V" shape that is slightly smaller than the "V" shape of the bottom.

Figure 4 also shows that through the 6 selected representative α angles of the V-shaped undulating bottom, the spatial distributions of the wave amplitudes in the focusing areas owning to Bragg resonance are symmetric about the center of the V-shaped undulating bottom, and the largest wave amplitudes are along the axis of symmetry. By comparison, it is found that the wave-focusing intensity first increases and then weakens with the decreasing V-shaped undulating bottom angles $\alpha = 178.21°$, $168.40°$, $162.24°$, $155.32°$, $148.58°$, and $140.47°$. When α is about $162.24°$, there are significant focusing areas and focal points, and the focusing intensity at this angle is the strongest.

4.1.2. Quantitative Analysis of Wave Amplitudes in the Focusing Areas for Regular Waves

To further quantitatively analyze and compare wave amplitude focusing characteristics at different angles α, the wave amplitude increases $A_{\max}/A_0$ are classified and counted. In addition, under the four levels of $A_{\max}/A_0 > 1$, $A_{\max}/A_0 \geq 1.5$, $A_{\max}/A_0 \geq 2$, and $A_{\max}/A_0 \geq 2.5$, the occurrence frequencies of the values of $A_{\max}/A_0$ in the whole simulation range are calculated and expressed respectively in the form of $P(A_{\max}/A_0 > 1)$, $P(A_{\max}/A_0 \geq 1.5)$, $P(A_{\max}/A_0 \geq 2)$ and $P(A_{\max}/A_0 \geq 2.5)$, respectively. The frequency formula of each level is as follows:

$$P(A_{\max}/A_0 \geq C) = \frac{N(A_{\max}/A_0 \geq C)}{N_x \times N_y} \tag{5}$$

where C is the specified level constant; $N(A_{\max}/A_0 \geq C)$ is the number of nodes corresponding to $A_{\max}/A_0 \geq C$; N_x and N_y are the numbers of nodes on the x and y-axis in the whole simulation range respectively. In this section, C is assigned 1, 1.5, 2, and 2.5 respectively, $N_x = N_y = 512$.

For example, $\alpha = 178.21°$, $176.42°$, $174.63°$, $168.40°$, $167.52°$, $162.24°$, $160.50°$, $155.32°$, $151.93°$, $148.58°$, and $140.47°$ are selected for comparison. As there is no wave-focusing effect when $\alpha = 90°$, it is not analyzed. Figure 5 shows the level-frequency statistical diagrams of wave amplitude changes at different angles α.

In Figure 5a, it shows that a series of nodes appear in the simulation range at all selected angles, but $P(A_{\max}/A_0 > 1)$ and $P(A_{\max}/A_0 \geq 1.5)$ gradually decrease with the decrease of angles α. $P(A_{\max}/A_0 > 1)$ decreases from 21.49% to 15.74%, while $P(A_{\max}/A_0 \geq 1.5)$ decreases from 5.32% to 0. It can be seen that the focusing area is decreasing. That is because the area of Bragg resonance decreases with the decrease of the angle α, and then the wave height focusing areas also reduce. In addition, that is why the range of angles α of V-shaped undulating bottom is $90° \leq \alpha \leq 180°$.

However, the wave-focusing effects are not only judged by the wave height in the focusing areas but also based on the overall amplitude increases of nodes in the wave height focusing areas, i.e., the values of $A_{\max}/A_0$. Therefore, Figure 5b further shows the frequency statistic results of $P(A_{\max}/A_0 \geq 2)$ and $P(A_{\max}/A_0 \geq 2.5)$. It shows that both $P(A_{\max}/A_0 \geq 2)$ and $P(A_{\max}/A_0 \geq 2.5)$ increase first and then decrease with the decrease of angles α. Moreover, when $160° < \alpha < 168°$, the overall value of $P(A_{\max}/A_0 \geq 2)$ is larger. That is to say, the total number of nodes of $A_{\max}/A_0 \geq 2$ is the largest. Meanwhile, when $\alpha = 162.24°$, the values of $P(A_{\max}/A_0 \geq 2)$ and $P(A_{\max}/A_0 \geq 2.5)$ are the largest in the simulation range, which implies that the number of nodes corresponding to $A_{\max}/A_0 \geq 2$ and $A_{\max}/A_0 \geq 2.5$ is the most, and the wave-focusing effect is the strongest. The spatial distribution of wave amplitude change shows that when $\alpha = 162.24°$, the scope of wave height focusing areas are relatively concentrated, the wave amplitude increases are the largest as a whole, and the wave-focusing effect is the best.

In conclusion, $\alpha = 162.24°$ is considered to be the optimal angle for V-shaped undulating bottom.

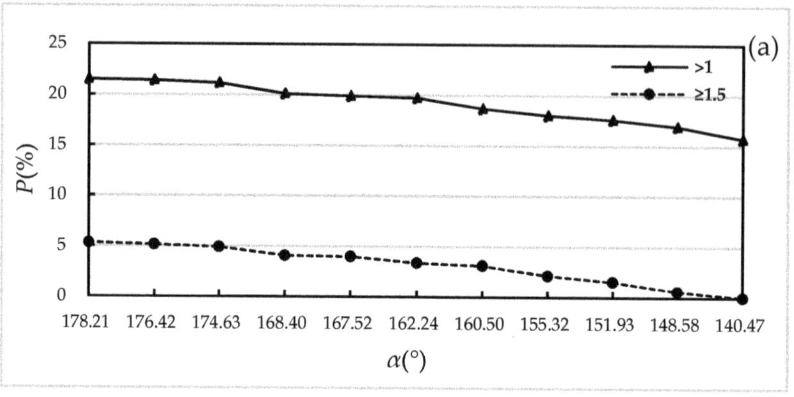

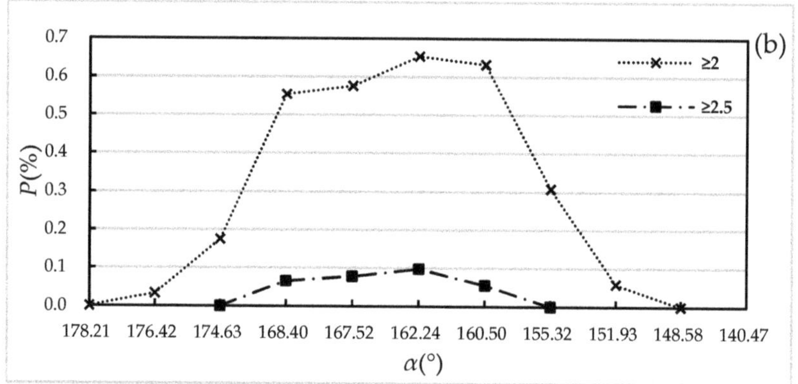

Figure 5. Level-frequency statistical diagrams of wave amplitude changes under different angles α: (a) $A_{max}/A_0 > 1$ and $A_{max}/A_0 \geq 1.5$; (b) $A_{max}/A_0 \geq 2$ and $A_{max}/A_0 \geq 2.5$.

4.2. Bragg Resonance Focusing Characteristics of Random Waves for V-Shaped Undulating Bottom

4.2.1. The Gaussian Spectrum

The Gaussian spectrum is a symmetric spectrum, and its spectrum pattern is simple and regular. The expression of the Gaussian spectrum requires only 3 parameters, compared with the Jonswap formula, which requires 5 parameters. Therefore, it is more convenient to use the Gaussian spectrum to calculate. In this paper, the two-dimensional Gaussian spectrum pattern is selected, and the expression is shown in Equation (6).

$$S(k_i) = \frac{\eta^2}{\sqrt{2\pi}\sigma} \exp(-\frac{(k_i - k_0)^2}{2\sigma^2}) \tag{6}$$

where: Gaussian spectrum expression $S(k_i)$, the standard deviation of the height of the wave surface $\eta = (\int S(k)dk)^{1/2} = \sqrt{m_0}$, zero-order spectrum moment m_0, spectral peak wave number k_0, spectral width parameter σ, value range of the i-th wave number $0 \leq k_i/dk \leq N/4$, the node number in the simulation range N.

The initial wave steepness and spectral width of the random wave are changed by controlling η and σ respectively. The initial wave steepness ε_0, initial amplitude a_0, and spectral width B_0 are calculated according to Equations (7)–(9) respectively.

$$\varepsilon_0 = 2\eta k_0 \tag{7}$$

$$a_0 = \varepsilon_0/k_0 = 2\eta = 2\sqrt{m_0} \tag{8}$$

$$B_0 = \sigma/k_0 \tag{9}$$

The amplitude of the *i*-th wave can be calculated by Equation (10).

$$a_i = \sqrt{2S(k_i)dk} \tag{10}$$

The initial wave surface and potential function can be obtained from Equation (11).

$$\left. \begin{array}{l} \eta(x,0) = \sum a_i \cos(k_i x + \theta_i) \\ \phi^s(x,0) = \sum \frac{g a_i}{\omega_i} \sin(k_i x + \theta_i) \end{array} \right\} \tag{11}$$

where: the initial wave surface $\eta(x,0)$, the initial potential function $\phi^s(x,0)$, the initial phase of the *i*-th wave generated randomly through the program θ_i, the gravitational acceleration g, the circular frequency of the *i*-th wave ω_i, according to the dispersion relation $\omega^2 = gk\tanh kh$ (water depth h) in the finite water depth to calculate the k_i.

Assuming that the initial wave surface and the initial potential function along the *y*-axis are equal, the two-dimension random wave field is extended into a three-dimension random wave field along the *y*-axis.

4.2.2. Evolution Characteristics of Random Waves on Flat Bottom

The Gaussian spectrum is used to generate the initial wave field. The initial phases of each simulated wave field are randomly generated. Different initial phases, dispersion relation, and wave modulation instability will cause uneven wave height distributions along the *x*-axis. Therefore, the evolution characteristics of random waves on the flat bottom are first studied in this study. The distributions of significant wave heights H_s along the *x*-axis under three random initial phases in the wave order $M = 3$ are shown in Figure 6. It shows that the distributions of significant wave heights along the *x*-axis are greatly different with different initial phases. Here, the node number along the *x*-axis is 512, and 16 nodes represent one wavelength of incident free surface.

As the number of simulation increases, the significant wave heights H_s under different random initial phases are averaged. The results are shown in Figure 7.

With the increase of the simulation groups, the distributions of significant wave heights H_s along the *x*-axis become more uniform. When the number of simulation groups is more than 10, the significant wave heights are basically stable in the form of a horizontal line, which meets the requirements of analysis.

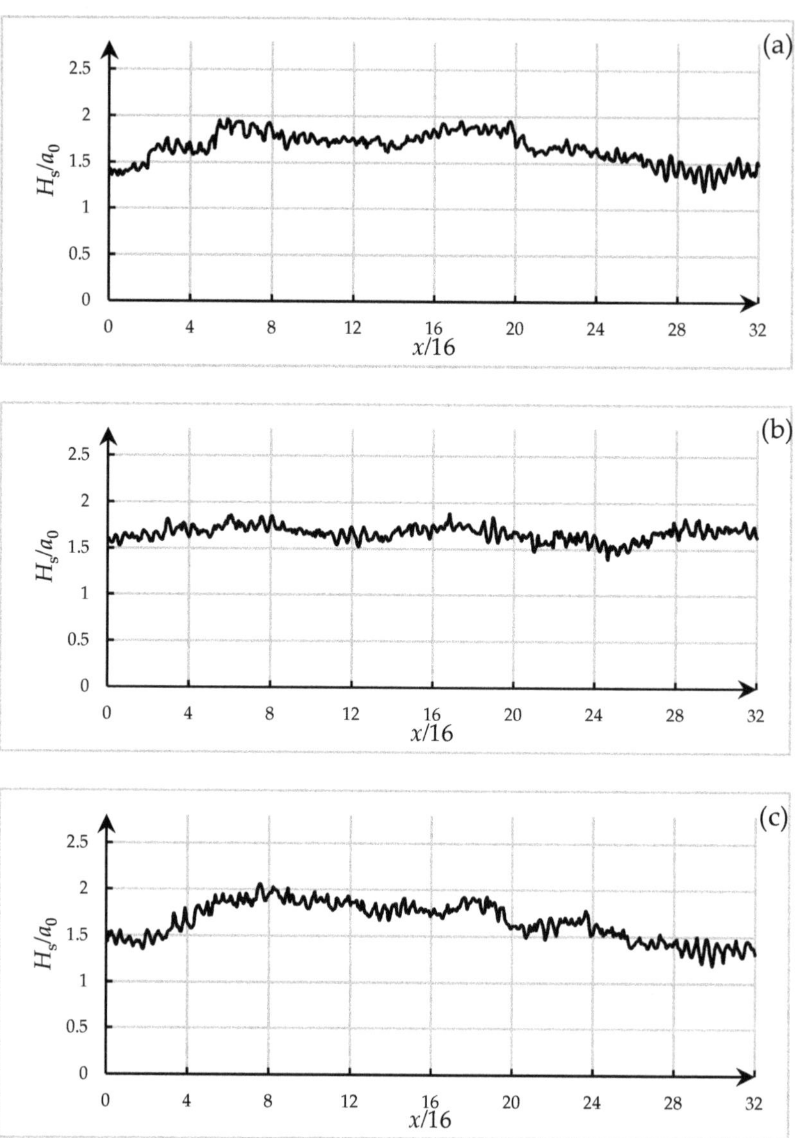

Figure 6. Distributions of significant wave heights H_s along the x-axis under three random initial phases for $M = 3$: (**a**) Initial phase one; (**b**) Initial phase two; (**c**) Initial phase three.

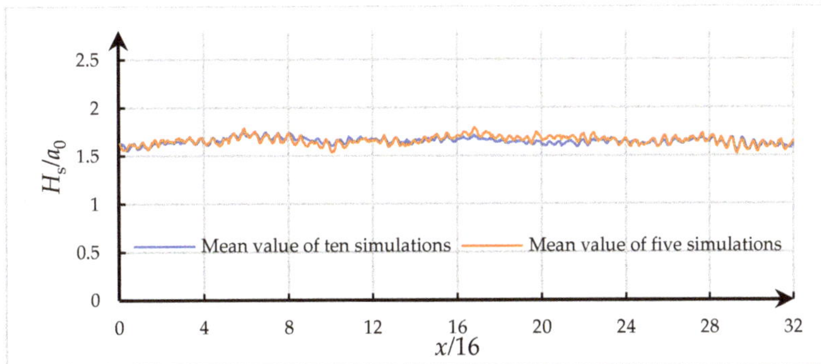

Figure 7. Distributions of average significant wave heights H_s along the x-axis under five and ten simulations for $M = 3$.

4.3. Evolution Characteristics of Random Waves on V-Shaped Undulating Bottom

BFI was proposed by Janssen [56], which is determined by wave steepness and spectrum width. The initial definition is shown in Equation (12), which is converted into the definition of wavenumber spectrum (14) by combining with Equation (13):

$$BFI = \sqrt{2}\varepsilon_0 / (2\Delta\omega / \omega_0) \tag{12}$$

$$2\Delta\omega / \omega_0 = \Delta k / k_0 \tag{13}$$

$$BFI = \sqrt{2}\varepsilon_0 / B_0 \tag{14}$$

where: the circular frequency ω, the initial circular frequency ω_0, the wavenumber of incident random waves k, the wavenumber of initial incident random waves k_0, the spectrum width B, the initial spectrum width B_0, the initial wave steepness ε_0.

Janssen proposed that when *BFI* is greater than or equal to 1.0, satisfying the conditions for generation of modulation instability, the possibility of freak waves is increased. To eliminate the influence of wave modulation instability on the simulation results, the *BFI* is set at less than 1.0 in this study. Under different combinations of initial wave steepness and spectrum width, the values of *BFI* are listed in Table 4.

Table 4. The values of *BFI* under the settings of the Gaussian spectrum.

ε_0	B_0	*BFI*	ε_0	B_0	*BFI*
	0.1	0.990		0.1	0.849
	0.2	0.495		0.2	0.424
0.07	0.3	0.330	0.06	0.3	0.283
	0.4	0.247		0.4	0.212
ε_0	B_0	*BFI*	ε_0	B_0	*BFI*
	0.1	0.707		0.1	0.566
	0.2	0.354		0.2	0.283
0.05	0.3	0.236	0.04	0.3	0.189
	0.4	0.177		-	-

The calculated wave heights (H_{smax}) are compared with the incident wave height (H_{s0}). To examine the interactions between random waves and undulating bottom, the spatial distribution characteristics of wave heights due to Bragg resonance are analyzed. Analyzing the combinations of initial wave steepness and initial spectrum width, it is found that there is a good linear relationship between H_{smax}/H_{s0} and *BFI*, as shown in Figure 8. For *BFI*, in the range of 0.15–1.0, the values of H_{smax}/H_{s0} linearly increase with the increase of *BFI*, so the fitting formula is shown in Equation (15):

$$H_{\text{smax}}/H_{s0} = 0.39 \times BFI + 1.09 \tag{15}$$

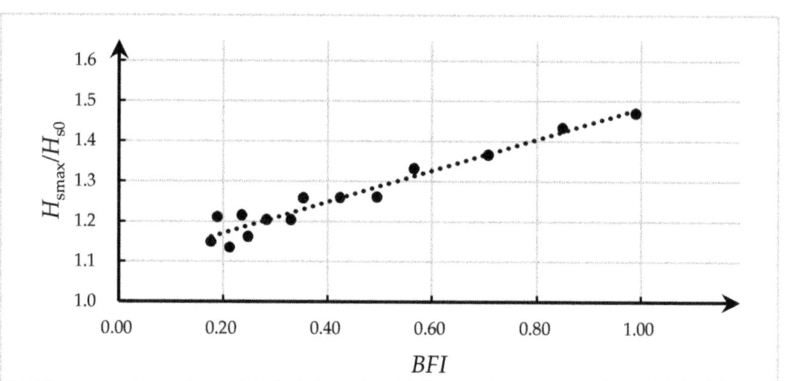

Figure 8. The changing curve of wave height ratio H_{smax}/H_{s0} at the focal points as *BFI* changes.

The goodness of fit (R^2) in Equation (15) can reach 0.94, indicating that the formula fits well. It is speculated that the calculation results will be more consistent with the line with the increase of the simulation groups. When the value of *BFI* is more than 0.4, there is a linear relationship between H_{smax}/H_{s0} and *BFI*; whereas the value of *BFI* is less than 0.4, H_{smax}/H_{s0} and *BFI* are weakly linear relation. When the *BFI* factor is small, in other words, the initial wave steepness of a given spectrum width is small, which indicates that the nonlinear wave interaction is weak. However, a given initial wave steepness has a larger spectrum width, indicating that the frequency bandwidth of wave height distribution is larger. In summary, the smaller the *BFI* factor is, the smaller H_{smax}/H_{s0} is; *BFI* and H_{smax}/H_{s0} are becoming more independent, when the incident wave nonlinearity is low.

5. Conclusions

In this study, the V-shaped undulating bottom can be regarded as the combination of two parts of oblique bottom undulations relative to the incident waves. The spatial distributions of the wave heights in the focusing areas owning to Bragg resonance of regular wave-bottom interactions are symmetric about the center of the V-shaped undulating bottom, and the largest wave height is distributed along the *x*-axis of symmetry. From these two aspects of spatial distributions and quantitative analysis, it is found that when α is about 162.24°, obvious focusing areas and focal points appear, and the focusing intensity at this angle is the strongest. By optimizing α, the wave-focusing effect is improved. When 145° < α < 180°, the wave amplitude increase at the focal points can reach more than 1.81 times of the initial incident wave amplitude, which is higher than that of the focal points when α = 180°. In particular, when α = 162.24°, the wave-focusing effect is the best and the maximum amplitude increases by 2.87 times of the initial incident wave amplitude. When α < 145°, the maximum amplitude increase of the focal points is less than that of α = 180°. When α = 90°, the maximum amplitude of the focal points is unchanged, i.e., wave energy focusing does not occur.

Therefore, the V-shaped undulating bottom at the angle of 162.24° is selected to study the random wave-bottom interactions characteristics. It is found that there is a good linear relationship between H_{smax}/H_{s0} and *BFI*, which combines wave steepness and spectrum width. For *BFI*, in the range of 0.15–1.0, the values of H_{smax}/H_{s0} linearly increase with the increase of *BFI*.

Author Contributions: Formal analysis, J.S.; investigation, S.X.; methodology: J.T.; software, H.Z., J.T. and J.S.; validation, S.X.; writing—original draft preparation, H.Z.; writing—review and editing, A.T. All authors have read and agreed to the published version of the manuscript.

Funding: This research is supported financially by the National Key Research and Development Program of China (2020YFD0900701) and the National Natural Science Foundation of China (U1706230).

Institutional Review Board Statement: Not applicable.

Informed Consent Statement: Not applicable.

Data Availability Statement: Not applicable.

Conflicts of Interest: The authors declare no conflict of interest.

References

1. Ahamed, R.; McKee, K.; Howard, I. Advancements of wave energy converters based on power take off (PTO) systems: A review. *Ocean Eng.* **2020**, *204*, 107248. [CrossRef]
2. Hayward, J.; Behrens, S.; McGarry, S.; Osman, P. Economic modelling of the potential of wave energy. *Renew. Energy* **2012**, *48*, 238–250. [CrossRef]
3. Mustapa, M.A.; Yaakob, O.; Ahmed, Y.M.; Rheem, C.K.; Koh, K.; Adnan, F.A. Wave energy device and breakwater integration: A review. *Renew. Sustain. Energy Rev.* **2017**, *77*, 43–58. [CrossRef]
4. Rahm, M.; Svensson, O.; Boström, C.; Waters, R.; Leijon, M. Experimental results from the operation of aggregated wave energy converters. *IET Renew. Power Gener.* **2012**, *6*, 149–160. [CrossRef]
5. Bragg, W.H. X-rays and crystalline structure. *Science* **1914**, *40*, 795–802. [CrossRef] [PubMed]
6. Davies, A.G. The reflection of wave energy by undulations on the seabed. *Dyn. Atmos. Ocean.* **1982**, *6*, 207–232. [CrossRef]
7. Heathershaw, A.D. Seabed-wave resonance and sand bar growth. *Nature* **1982**, *296*, 343–345. [CrossRef]
8. Davies, A.G.; Heathershaw, A.D. Surface-wave propagation over sinusoidally varying topography. *J. Fluid Mech.* **1984**, *144*, 419–443. [CrossRef]
9. Mitra, A.; Greenberg, M.D. Slow interactions of gravity waves and a corrugated seabed. *Trans. ASME J. Appl. Mech.* **1984**, *51*, 251–255. [CrossRef]
10. Mei, C.C. Resonant reflection of surface water waves by periodic sandbars. *J. Fluid Mech.* **1985**, *152*, 315–335. [CrossRef]
11. Kirby, J.T. A general wave equation for waves over rippled beds. *J. Fluid Mech.* **1986**, *162*, 171–186. [CrossRef]
12. Bailard, J.A.; Devries, J.W.; Kirby, J.T. Considerations in using Bragg reflection for storm erosion protection. *J. Waterw. Port Coast. Ocean Eng.* **1992**, *118*, 62–74. [CrossRef]
13. Bailard, J.A.; Devries, J.; Kirby, J.T.; Guza, R.T. Bragg reflection breakwater: A new shore protection method? In Proceedings of the 22nd International Conference on Coastal Engineering Proceedings, Delft, The Netherlands, 2–6 July 1990; pp. 1702–1715.
14. Kirby, J.T.; Anton, J.P. Bragg reflection of waves by artificial bars. In Proceedings of the 22nd International Conference on Coastal Engineering Proceedings, Delft, The Netherlands, 2–6 July 1990; pp. 757–768.
15. O'Hare, T.J.; Davies, A.G. Sand bar evolution beneath partially-standing waves: Laboratory experiments and model simulations. *Cont. Shelf Res.* **1993**, *13*, 1149–1181. [CrossRef]
16. Yu, J.; Mei, C.C. Do longshore bars shelter the shore? *J. Fluid Mech.* **2020**, *404*, 251–268. [CrossRef]
17. Howard, L.N.; Yu, J. Normal modes of a rectangular tank with corrugated bottom. *J. Fluid Mech.* **2007**, *593*, 209–234. [CrossRef]
18. Weidman, P.D.; Herczyński, A.; Yu, J.; Howard, L.N. Experiments on standing waves in a rectangular tank with a corrugated bed. *J. Fluid Mech.* **2015**, *777*, 122–150. [CrossRef]
19. Mei, C.C.; Hara, T.; Naciri, M. Note on Bragg scattering of water waves by parallel bars on the seabed. *J. Fluid Mech.* **1988**, *186*, 147–162. [CrossRef]
20. Liu, Y.; Yue, D.K.P. On generalized Bragg scattering of surface waves by bottom ripples. *J. Fluid Mech.* **1998**, *356*, 297–326. [CrossRef]
21. Alam, M.R.; Liu, Y.; Yue, D.K.P. Oblique sub- and super-harmonic Bragg resonance of surface waves by bottom ripples. *J. Fluid Mech.* **2010**, *643*, 437–447. [CrossRef]
22. Couston, L.A.; Jalali, M.A.; Alam, M.R. Shore protection by oblique seabed bars. *J. Fluid Mech.* **2017**, *815*, 481–510. [CrossRef]
23. Alam, M.R.; Liu, Y.; Yue, D.K.P. Attenuation of short surface waves by the sea floor via nonlinear sub-harmonic interaction. *J. Fluid Mech.* **2011**, *689*, 529–540. [CrossRef]
24. Zhang, J.; Benoit, M. Effect of finite amplitude of bottom corrugations on Fabry-Perot resonance of water waves. *Phys. Rev. E* **2019**, *99*, 053109. [CrossRef] [PubMed]
25. Porter, D.; Staziker, D.J. Extensions of the mild-slope equation. *J. Fluid Mech.* **1995**, *300*, 367–382. [CrossRef]
26. Athanassoulis, G.; Belibassakis, K.A. A consistent coupled-mode theory for the propagation of small-amplitude water waves over variable bathymetry regions. *J. Fluid Mech.* **1999**, *389*, 275–301. [CrossRef]
27. Belibassakis, K.; Athanassoulis, G.; Gerostathis, T. A coupled-mode model for the refraction–diffraction of linear waves over steep three-dimensional bathymetry. *Appl. Ocean Res.* **2001**, *23*, 319–336. [CrossRef]

28. O'Hare, T.J.; Davies, A.G. A new model for surface wave propagation over undulating topography. *Coast. Eng.* **1992**, *18*, 251–266. [CrossRef]

29. Seo, S.N. Transfer matrix of linear water wave scattering over a stepwise bottom. *Coast. Eng.* **2014**, *88*, 33–42. [CrossRef]

30. Madsen, P.A.; Fuhrman, D.R.; Wang, B. A Boussinesq-type method for fully nonlinear waves interacting with a rapidly varying bathymetry. *Coast. Eng.* **2006**, *53*, 487–504. [CrossRef]

31. Dommermuth, D.G.; Yue, D.K.P. A high-order spectral method for the study of nonlinear gravity waves. *J. Fluid Mech.* **1987**, *184*, 267–288. [CrossRef]

32. Guazzelli, E.; Rey, V.; Belzons, M. Higher-order Bragg reflection of gravity surface waves by periodic beds. *J. Fluid Mech.* **1992**, *245*, 301–317. [CrossRef]

33. Chang, H.K.; Liou, J.C. Long wave reflection from submerged trapezoidal breakwaters. *Ocean Eng.* **2007**, *34*, 185–191. [CrossRef]

34. Wen, C.C.; Tsai, L.H. Numerical simulation of Bragg reflection based on linear waves propagation over a series of rectangular seabed. *China Ocean Eng.* **2008**, *22*, 71–86.

35. Zhang, J.S.; Jeng, D.S.; Liu, P.F.; Zhang, C.; Zhang, Y. Response of a porous seabed to water waves over permeable submerged breakwaters with Bragg reflection. *Ocean Eng.* **2012**, *43*, 1–12. [CrossRef]

36. Ouyang, H.T.; Chen, K.H.; Tsai, C.M. Wave characteristics of Bragg reflections from a train of submerged bottom breakwaters. *J. Hydro-Environ. Res.* **2016**, *11*, 91–100. [CrossRef]

37. Yueh, C.Y.; Chuang, S.H.; Wen, C.C. Bragg reflection of water waves due to submerged wavy plate breakwater. *J. Hydro-Environ. Res.* **2018**, *21*, 52–59. [CrossRef]

38. Ning, D.; Chen, L.; Zhao, M.; Teng, B. Experimental and numerical investigation of the hydrodynamic characteristics of submerged breakwaters in waves. *J. Coast. Res.* **2016**, *32*, 800–813. [CrossRef]

39. Shih, R.S.; Weng, W.K. Experimental determination of the performance characteristics of an undulating submerged obstacle. *Ships Offshore Struct.* **2016**, *11*, 129–141. [CrossRef]

40. Liu, Y.; Li, H.-J.; Zhu, L. Bragg reflection of water waves by multiple submerged semi-circular breakwaters. *Appl. Ocean Res.* **2016**, *56*, 67–78. [CrossRef]

41. Zheng, J.; Yao, Y.; Chen, S.; Chen, S.; Zhang, Q. Laboratory study on wave-induced setup and wave-driven current in a 2DH reef-lagoon-channel system. *Coast. Eng.* **2020**, *162*, 103772. [CrossRef]

42. Elandt, R.B.; Couston, L.A.; Lambert, R.A.; Alam, M.R. Bragg Resonance of Gravity Waves and Ocean Renewable Energy. *Integr. Syst. Innov. Appl.* **2015**, 211–226. [CrossRef]

43. Tao, A.F.; Yan, J.; Wang, Y.; Zheng, J.H.; Fan, J.; Qin, C. Wave power focusing due to the Bragg resonance. *China Ocean Eng.* **2017**, *31*, 458–465. [CrossRef]

44. Zhang, C.; Ning, D. Hydrodynamic study of a novel breakwater with parabolic openings for wave energy harvest. *Ocean Eng.* **2019**, *182*, 540–551. [CrossRef]

45. Phillips, O.M. On the dynamics of unsteady gravity waves of finite amplitude. I: The elementary interactions. *J. Fluid Mech.* **1960**, *9*, 193–217. [CrossRef]

46. Wu, G.Y. Direct Simulation and Deterministic Prediction of Large Scale Nonlinear Ocean Wave Field. Ph.D. Thesis, Massachusetts Institute of Technology, Cambridge, MA, USA, 2004.

47. Zhao, X.Z.; Sun, Z.C.; Liang, S.X. Efficient focusing models for generation of freak waves. *China Ocean Eng.* **2009**, *23*, 429–440.

48. Tao, A.-F.; Zheng, J.-H.; Mee, M.S.; Chen, B.-T. Re-study on recurrence period of Stokes wave train with High Order Spectral method. *China Ocean Eng.* **2011**, *25*, 679–686. [CrossRef]

49. Seiffert, B.R.; Ducrozet, G.; Bonnefoy, F. Simulation of breaking waves using the high-order spectral method with laboratory experiments: Wave-breaking onset. *Ocean Model.* **2017**, *119*, 94–104. [CrossRef]

50. Seiffert, B.R.; Ducrozet, G. Simulation of breaking waves using the high-order spectral method with laboratory experiments: Wave-breaking energy dissipation. *Ocean Dyn.* **2018**, *68*, 65–89. [CrossRef]

51. Song, J.; Zhuang, Y.; Wan, D. New wave spectrums models developed based on HOS method. In Proceedings of the 28th International Ocean and Polar Engineering Conference, Sapporo, Japan, 10–15 June 2018; pp. 524–531.

52. Guo, Q.; Alam, M.R. Prediction of oceanic rogue waves through tracking energy fluxes. In Proceedings of the 36th International Conference on Ocean, Offshore and Arctic Engineering, Trondheim, Norway, 25–30 June 2017.

53. Gouin, M.; Ducrozet, G.; Ferrant, P. Propagation of 3D nonlinear waves over an elliptical mound with a High-Order Spectral. *Eur. J. Mech. B/Fluids* **2017**, *63*, 9–24. [CrossRef]

54. Fan, J.; Zheng, J.; Tao, A.; Liu, Y. Upstream-propagating waves induced by steady current over a rippled bottom: Theory and experimental observation. *J. Fluid Mech.* **2021**, *910*, A49. [CrossRef]

55. Zakharov, V. Stability of periodic waves of finite amplitude on the surface of a deep fluid. *J. Appl. Mech. Tech. Phys.* **1968**, *9*, 86–94. [CrossRef]

56. Janssen, P. Nonlinear four-wave interactions and freak waves. *J. Phys. Oceanogr.* **2003**, *33*, 863–884. [CrossRef]

Journal of
Marine Science and Engineering

Article

Deriving the 100-Year Total Water Level around the Coast of Corsica by Combining Trivariate Extreme Value Analysis and Coastal Hydrodynamic Models

Jessie Louisor [1,*], Jérémy Rohmer [1], Thomas Bulteau [2], Faïza Boulahya [1], Rodrigo Pedreros [1], Aurélie Maspataud [2] and Julie Mugica [3]

[1] BRGM, 3 Av. C. Guillemin BP 36009, CEDEX 2, 45060 Orléans, France; j.rohmer@brgm.fr (J.R.); f.boulahya@brgm.fr (F.B.); r.pedreros@brgm.fr (R.P.)
[2] BRGM, 24 Avenue Léonard de Vinci, 33600 Pessac, France; t.bulteau@brgm.fr (T.B.); a.maspataud@brgm.fr (A.M.)
[3] BRGM, Immeuble Agostini, Z.I. de Furiani, 20600 Bastia, France; j.mugica@brgm.fr
*   Correspondence: j.louisor@brgm.fr

Abstract: As low-lying coastal areas can be impacted by flooding caused by dynamic components that are dependent on each other (wind, waves, water levels—tide, atmospheric surge, currents), the analysis of the return period of a single component is not representative of the return period of the total water level at the coast. It is important to assess a joint return period of all the components. Based on a semiparametric multivariate extreme value analysis, we determined the joint probabilities that significant wave heights (Hs), wind intensity at 10 m above the ground (U), and still water level (SWL) exceeded jointly imposed thresholds all along the Corsica Island coasts (Mediterranean Sea). We also considered the covariate peak direction (Dp), the peak period (Tp), and the wind direction (Du). Here, we focus on providing extreme scenarios to populate coastal hydrodynamic models, SWAN and SWASH-2DH, in order to compute the 100-year total water level (100y-TWL) all along the coasts. We show how the proposed multivariate extreme value analysis can help to more accurately define low-lying zones potentially exposed to coastal flooding, especially in Corsica where a unique value of 2 m was taken into account in previous studies. The computed 100y-TWL values are between 1 m along the eastern coasts and a maximum of 1.8 m on the western coast. The calculated values are also below the 2.4 m threshold recommended when considering the sea level rise (SLR). This highlights the added value of performing a full integration of extreme offshore conditions, together with their dependence on hydrodynamic simulations for screening out the coastal areas potentially exposed to flooding.

Keywords: multivariate extreme value; coastal modeling; SWAN; SWASH-2DH; Corsica; return level; total water level

Citation: Louisor, J.; Rohmer, J.; Bulteau, T.; Boulahya, F.; Pedreros, R.; Maspataud, A.; Mugica, J. Deriving the 100-Year Total Water Level around the Coast of Corsica by Combining Trivariate Extreme Value Analysis and Coastal Hydrodynamic Models. *J. Mar. Sci. Eng.* **2021**, *9*, 1347. https://doi.org/10.3390/jmse9121347

Academic Editors: Wei-Bo Chen, Shih-Chun Hsiao and Wen-Son Chiang

Received: 30 September 2021
Accepted: 24 November 2021
Published: 30 November 2021

Publisher's Note: MDPI stays neutral with regard to jurisdictional claims in published maps and institutional affiliations.

1. Introduction

Coastal flooding often occurs because of several dynamic components that are dependent on each other (wind, waves, water levels—tide, atmospheric surge, currents); hence, the analysis of the return period of a single component is not representative of the return period of the total water level at the coast [1]. As such, it is important to estimate the joint return period, taking into account the dependency between all of the components. Marcos et al. [2] even show that return periods of extreme sea levels are underestimated (by a factor of 2 or higher) in 30% of the coasts (at global scale), if the dependency is neglected.

The dependency can also be based on a spatial linkage assumption. Galiatsatou and Prinos [3] took into account the degree of spatial dependence between the sites, in order to estimates extreme storm surges along the Dutch coast on the North Sea.

Some authors considered extreme quantiles of metocean data (significant wave height, wave period, and wind speed) estimated jointly, using a subset exceeding appropriately

J. Mar. Sci. Eng. **2021**, *9*, 1347. https://doi.org/10.3390/jmse9121347   https://www.mdpi.com/journal/jmse

defined thresholds [4]. Moreover, several joint probability approaches applied to sea conditions have been used by [5–9], for example. The reader might also be interested in the review proposed by [10]. One of the limitations encountered within the classical joint probability approaches are related to the restrictive assumptions regarding the dependence structure when extrapolating to extremes. Idier et al. [11] mentioned that one of the difficulties with such approaches could be the number of offshore parameters, which can complicates the detection of the critical conditions. Heffernan and Town [12] developed a semiparametric approach (hereinafter called H&T04 methods), which overcomes the limitations imposed by dimensionality. This approach was first applied to air pollution data. Since then, several authors applied the H&T04 methods to metocean offshore conditions: [13–16].

In the present study, we are interested in coastal flooding all around Corsica Island in the Mediterranean Sea. To our best knowledge, a multivariate extreme value analysis has never been implemented along these coasts, though some studies undertook statistical analysis over the Mediterranean Sea or around Corsica, e.g., [17]. In particular, in 2004, the Western European Union funded a Wind and Wave Atlas of the Mediterranean Sea, providing both bivariate and univariate statistics, and spatial distribution of statistical quantities for wind and wave parameters [18]. This atlas was built 20 years ago with a 10-year dataset consisting of less than 1000 data points (spatial resolution of the outputs ~50 km), distributed throughout the whole Mediterranean Sea using dynamical models which have been updated since the beginning of the 21st century. Other initiatives in this basin should be noted. For example, the WaveForUs (Wave climate and coastal circulation Forecasts for public Use) platform provides 3-day meteorological and sea state forecasts in the Mediterranean Sea since 2013 with a 0.15° spatial resolution [19]. To date, these archives are too short to be used as such in local coastal flooding assessments.

To be more accurate at local scale, a solution could be to apply dynamical downscaling using a complex nested model chain. However, this is particularly delicate in the Mediterranean Sea because of specific issues mentioned in [20]: especially the orography (for winds and pressure), and the complex bathymetry, and limited fetch extension (for waves). Several authors proposed an alternative in order to reduce the complexity and the computational cost of such a numerical chain [21,22]. Indeed, based on the forcing event selection, which consist in: (1) determining the joint probabilities that significant wave heights Hs, wind intensity at 10 m above the ground U, and still water level SWL exceed jointly imposed thresholds; (2) defining the extreme combinations; (3) populating the coastal hydrodynamic models at a local scale with the extreme combinations as input. For instance, the authors in reference [23] used dozens of scenarios of offshore forcing conditions (significant wave height and total water level) associated with the same joint probability of exceedance (here with a 100-year return period) as inputs of their modeling chain. The appealing feature of this approach is its computational simplicity and efficiency, because it is only based on a limited number of scenarios, i.e., on a limited number of long-running numerical simulations.

From an operational point of view, our primary motivation is based on the fact that, since 2013, a unique value of 2 m relative to the French national topographic reference (NGF) was taken into account to map entire zones potentially exposed to coastal flooding in Corsica (e.g., http://carto.geo-ide.application.developpement-durable.gouv.fr/429/risques_naturels_02a.map (accessed on 2 September 2020)). The reader may note that all data and results presented in this study are relative to the NGF. For the local stakeholders and coastal managers, it is important to assess if this 2 m value is relevant in their region [24]. The main purpose of the present study is to show how multivariate extreme value analysis can help to better define low-lying zones potentially exposed to coastal flooding.

In this paper, we have first described the site (location, and context), and data used in Section 2. We then exposed the methods applied to constitute extreme scenarios based on triplets (Hs; SWL; U) for significant wave height, still water level, and wind speed. The strategy used to propagate offshore conditions to the coastline has also been presented in this section. In Section 3, we have applied the proposed strategy all around Corsica, considering extreme scenarios under both, current and climate change conditions. The results are presented in Section 4. Finally, a discussion of our results, and the limitations of the method are presented in Section 5.

2. Materials and Methods

2.1. Site Description and Data Overview

The study site is Corsica Island, located in the Mediterranean Sea (Figure 1). Since 2013, the approach implemented for the whole French Mediterranean coastline consists in mapping coastal sectors whose altitude is below 2 m. These sectors are considered potentially exposed to coastal flooding. Indeed, in past studies, the 100-year return period event was computed using the Sète tide gauge located at least 400 km from the coasts studied (red symbol in Figure 1). Moreover, the analysis did not take into account the combination of wind, waves, and water level as a joint return period. For coastal management, it is necessary to assess if this 2 m value is valid in our study area.

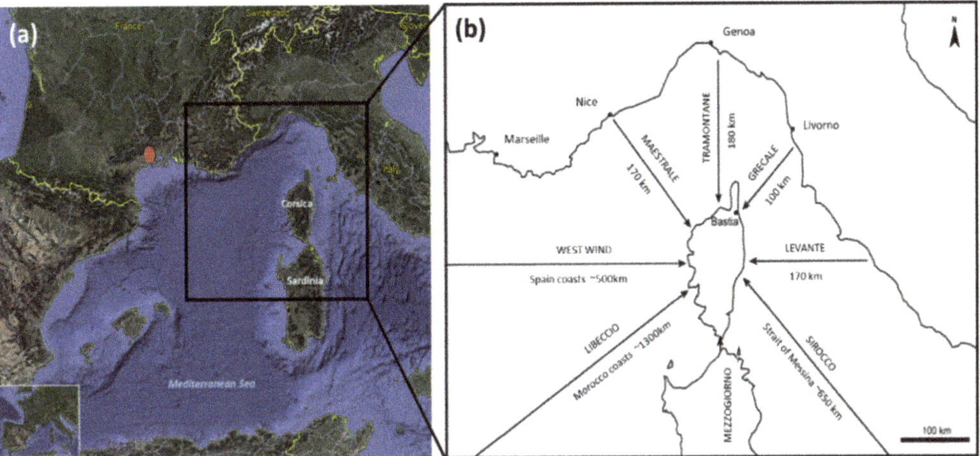

Figure 1. Location of Corsica Island and Sète (red symbol). (**a**) The map is adapted from Google Earth. (**b**) Wind provenance and crossed distance. The map is adapted from ROL Corse (Coastal Observatory Network of Corsica).

Figure 1b highlights that the island is under the influence of different wind regimes coming from far as the Libeccio (can cross more than 1000 km) or coming from closer as the Mezzogiorno (~10 km). The waves induced by the wind can be swell when they cross long distance or wind waves if they stay close to their location of generation. Because of the distance from other coasts, we can assume that the west coast of Corsica can be impacted by swells, whereas the south and east coast mostly by wind waves.

In this area, the tidal range is from 0.2 to 0.4 m as it is a microtidal regime [25]. For this reason, we consider a direct approach using the SWL instead of storm surges. Indeed, Haigh et al. [26] showed that for long return periods, when the ratio of the tidal to non-tidal component is lower than 2, direct and non-direct approaches present good agreement. Kergadallan [27] showed that this condition (<2) is observed for the French Mediterranean coasts.

In order to manage our statistical analysis, validate data, and run hydrodynamic simulations, we used different products listed at Table 1.

Table 1. Data used in the study, spatial resolution, temporal resolution, and sources.

	Variable	Product	Source	Spatial and Temporal Resolution
Offshore conditions for statistical analysis	Sea states: Hs, Tp, Dp	NWW3_MED	NOAA https://polar.ncep.noaa.gov/waves/hindcasts/nopp-phase2/ (accessed on 2 July 2019)	20 km 3 h
	Winds: U, DU	NWW3_MED	NOAA	20 km 3 h
	Still Water Level SWL	MARS_MED_BRGM hindcast	BRGM	10 km around Corsica 10 min
Topobathymetric data for numerical simulation	Bathymetry	HOMONIM	Shom https://data.shom.fr/ (accessed on 14 May 2019)	100 m
	Topography	RGE ALTI and Litto 3D	IGN https://www.geoportail.gouv.fr/donnees/rge-alti-sources (accessed on 14 May 2019)	1 m
In-situ data used for validation	Hs, Tp, Dp	Buoy time series	Candhis https://candhis.cerema.fr/ (accessed on 1 July 2019) and ISPRA http://dati.isprambiente.it/id/website/ronRmn/html (accessed on 2 July 2019)	1 h or 30 min
	U, DU	Station and buoy time series	Météo-France https://donneespubliques.meteofrance.fr/ (accessed on 3 July 2019) and ISPRA	10 min
	SWL	Tide gauge time series	Shom and CNES https://www.aviso.altimetry.fr/en/data/products/in-situ-products/tide-gauge.html (accessed on 1 July 2019)	10 min

The National Oceanic and Atmospheric Administration (NOAA) provides the NWW3 MED hindcast [28,29] consisting of sea states, simulated using the third generation wave model Wavewatch 3 (WW3) with the parametrization proposed by [30]. Offshore wind conditions are also provided.

The SWL hindcast was generated by the BRGM using the MARS-2DH model [31]. This dataset is available from 1979 to 2010 and gives an assessment of the regional hydrodynamics based on tidal components and meteorological reanalysis CFSR [32].

The bathymetry and topography used after in the nearshore numerical modeling chain were provided by the Shom [33] and IGN.

We used in-situ data from Candhis and ISPRA (Table 1) to correct the Hs in the NWW3_MED hindcast using a linear correction (see Appendix A). This correction reduces the bias and is relevant for both mean and Hs > 4 m. For extreme Hs, the bias reduction is ~1 m. In addition, we used the in-situ datasets provided by the Shom and CNES to validate the MARS_MED_ BRGM hindcast (Appendix A), and in-situ data coming from tide gauges provided by ISPRA and Météo-France in order to validate wind speed and direction.

2.2. Statistical and Numerical Strategy

The purpose was to determine triplets (Hs; SWL; U) of a 100-year joint exceedance return period and their covariates (Tp, Dp, Du) to populate coastal hydrodynamic models. In order to perform the statistical analysis, we used MATLAB® software. The statistical method was implemented using the WAFO toolbox [34] and developments by Guanche Garcia [35]. The different steps are illustrated in Figure 2a and exposed here:

1. Selection of events and data set constitution: build a sample with a large number of independent triplets (Hs; SWL; U);
2. Fitting of the marginal probability laws for each variable Hs, SWL, and U using the Generalized Pareto Distribution (GPD) with bootstrap simulation for Confidence Interval (CI) [36];
3. Dependency model adjustment:
   (a) Between the extreme variables Hs, SWL, U, and based on the semiparametric approach described by [12]. The readers may also consult [37] for detailed information and application of this approach;
   (b) Between Hs et Tp using the empirical conditional distribution of wave steepness (St) knowing Hs described by [38];
4. Running Monte-Carlo simulations using the marginal laws and dependency models of a very large number of combinations (Hs; SWL; U) and their covariates Tp, Dp, Du with the same statistical characteristics as the observed data;
5. Determine the triplets (Hs; SWL; U) of the 100-year joint exceedance return period;
6. Running SWAN (Simulating Waves Nearshore) and SWASH-2DH (Simulating WAves till SHore) models for 50 m resolution grids (in red in Figure 2b) forced with 100-year triplets + covariates Tp, Dp, Du + safety margins for water level (i.e., sea level rise SLR + uncertainties);
7. Dynamical downscaling by running SWAN model for the 10 m resolution nested boxes (in orange in Figure 2b). SWAN is here forced by the outputs of step 6;
8. Assess the total water level (TWL) at the shoreline adding contributions coming from the models: wave setup provided by SWAN and wind setup by SWASH-2DH.

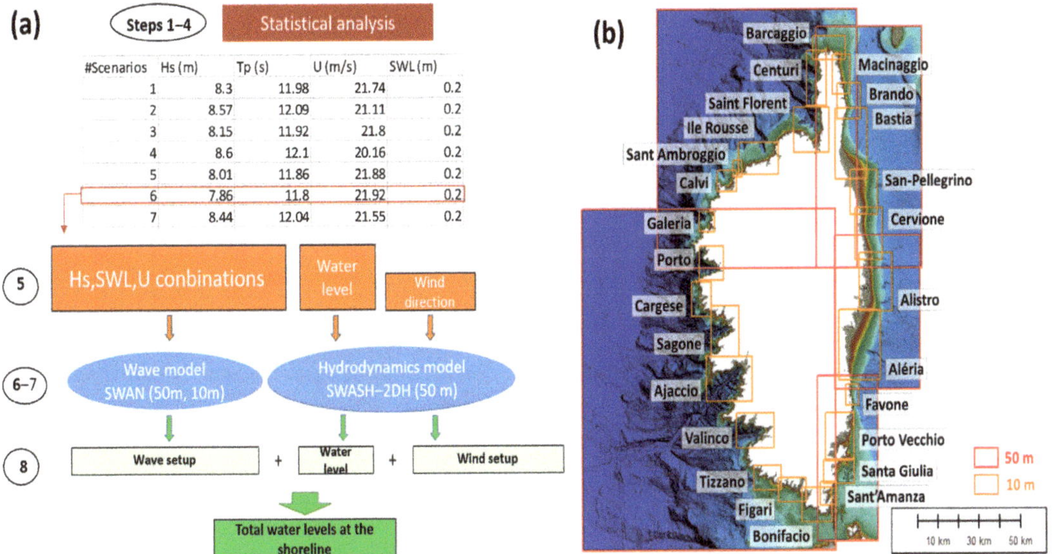

Figure 2. (a) Recap of our multivariate extreme value analysis. (b) Nested boxes used all around Corsica Island for the numerical simulations; 50 m resolution grids are in red and 10 m resolution boxes are in orange.

We applied the statistical analysis for points located in each of the red boxes shown in Figure 2b. The triplets obtained and their covariates finally forced the SWAN and SWASH-2DH models. For more information on the SWAN and SWASH-2DH open source models, the reader is invited to refer to [39–45].

Moreover, as mentioned at step 6, some safety margins are taken into account in the numerical simulations. Following the French regulation [46], the water level must account the Sea Level Rise (SLR) potentially due to climate change. To do so, 0.20 m is added to the water level of the 100-year event defined offshore. In order to consider a "2100" future scenario, 0.60 m has to be added to the 100-year return level defined offshore by following the French regulation guideline. The values recommended in the guideline are based on the Fifth Assessment Report (AR5), chapter 13 [47] published by the Intergovernmental Panel on Climate Change (IPCC). Note that the value of 0.60 m is in agreement with the mean SLR value estimated for the Mediterranean basin in [48,49], when considering the pessimistic scenario of Representative Concentration Pathway (RCP) 8.5. Moreover, [46] recommends taking into account uncertainties adding a margin of 0.25 m to the 100-year return level if all the uncertainties cannot be estimated. There are many sources of uncertainties, and even if many of them showed small-to-moderate degree of uncertainty (confidence interval given with marginals, correction on Hs), and that parametric tests regarding the statistical approach showed moderate impact as well (diagnosis on thresholds in H&T04), residual uncertainties remain. These residual uncertainties correspond to altimetric variations of Earth's crust, local vertical ground motions (e.g., due to groundwater pumping), impacts of remote contributions in a semi-enclosed basin such as the Mediterranean Sea or complex oceanic processes in the Strait of Gibraltar. To account for them, a margin of 0.25 m is applied following the French regulators' recommendations of [46]. We have considered two cases:

(i)   Present-day conditions: 100-year triplets/covariates + 0.2 m SLR + 0.25 m uncertainties;
(ii) "2100" future conditions: 100-year triplets/covariates + 0.6 m SLR + 0.25 m uncertainties.

3. Multivariate Extreme Value Analysis

In this section, we provide further implementation details on the steps 1–5 of the procedure described in Section 2.2, which aims at estimating the extreme offshore conditions to be used as forcing inputs of the SWAN-SWASH 2DH modeling chain (see Figure 2a). These steps are illustrated with the box around the Gulf of Porto (Figure 2b)

3.1. Selection of Events and Dataset Set Up (Step 1)

The first step of the procedure consists in selecting independent combinations (Hs; SWL; U). To do so, we first performed a block maxima selection to define each wave event (statistically independent) [50]. Indeed, based on a 40-year historical research in archives, press and in-situ data; and in consultation with the deconcentrated services of the state, we assumed that Hs is the primary driver of coastal flooding around the coast of Corsica. Following [23,51], we defined a block as a 3-day period with 1.5 days between each Hs peak in order to ensure the independence between each peak. Then, for each Hs peak, the SWL and U maxima were sought in a 12-h window centered on the Hs peak. This 12 h window allows the data of SWL to be scanned over approximately one tidal cycle. Each Hs peak is associated to the corresponding covariates, peak period Tp and peak direction Dp. Likewise, each value of U is associated with its direction Du. Thus, several combinations (Hs; Tp; Dp; SWL; U; Du) corresponding to 30 years of common data with ~100 events per year were selected in 7 different zones around Corsica.

3.2. Marginal Distributions (Step 2)

Figure 3 illustrates an example of the marginal probability distribution fitting $F_i(x)$ of each variable X_i for (Hs; SWL; U) using the GPD defined as:

$$F_i(x) = \begin{cases} \widetilde{F}_i(x) & x \leq u_i \\ 1 - \left(1 - \widetilde{F}_i(u_i)\right)\left[1 + \xi_i \frac{(x-u_i)}{\sigma_i}\right]_+^{-\frac{1}{\xi_i}} & x > u_i \end{cases} \tag{1}$$

where u_i is a high threshold to choose, σ_i and ξ_i are the GPD parameters.

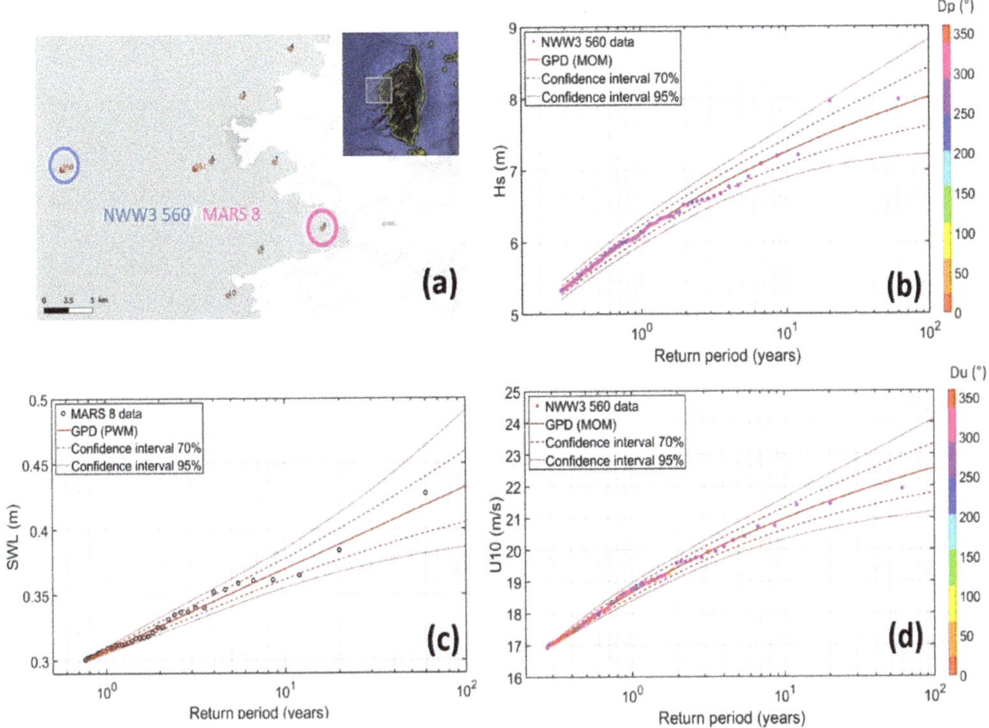

Figure 3. Example of the marginal distributions' fitting. (**a**) Location of the points considered. NWW3 560 provides U (m/s) and Hs (m), MARS 8 provides SWL (m). Generalized Pareto Distribution (GPD) and associated 70% (red dashed lines) and 95% (red thin lines) CI for (**b**) Hs, (**c**) SWL, and (**d**) U. Color bars give the directions Dp and Du (°).

The data points considered are located close to the Gulf of Porto (see Figure 3a). Here the point NWW3 560 provides U (m/s) and Hs (m), and the point MARS 8 gives the SWL (m). For SWL, the threshold u = 0.3 m, the 100-year return level is 0.43 m. The threshold is set to 5.32 m, and the 100-year Hs is 8.016 m for Hs, and set to 16.92 m/s for U leading to a 100-year U of 22.55 m/s. Different thresholds have been chosen for each zone around Corsica (see Appendix B). To do so, we applied several analysis of mean residual life plots and analysis of quantitative values resulting from statistical tests Chi-squared, Kolmogorov–Smirnov, or quantile-quantile graphs, based on [50] (not shown).

In addition, three methods for estimating the GPD parameters have been tested: the method of moments (MOM), the weighted method of moments (PWM), and the maximum likelihood method (MLE) [52]. Here (Figure 3b,d), the MOM was used to determine the parameters of the laws for Hs and U and the PWM method for SWL (Figure 3c), because these methods were the most conservative. Directions Dp and Du associated to Hs and U, respectively, are also reported in color. In this zone, the extremes are globally due to winds and waves with directions between 250 and 300°. Note that we chose to keep all directions Dp and Du, not only extremes coming from a given sector, because: (i) critical Dp for present day conditions are expected to change in the future due to future geomorphological changes on sandy coasts; (ii) only keeping data coming from a "given" sector can considerably reduce the sample size and increase the uncertainty in the statistical methods (marginals and dependence model). For instance, only 17% of data are kept if the dataset is split following directions between 200° and 250° for the NWW3 465/MARS 26 dataset (see Appendix C). Finally, the uncertainty associated with the distribution fitting is translated into the form of confidence intervals given using the parametric bootstrap [53]. Uncertainties related to the duration of the dataset may exist, but in our case, we have reduced them by using time series from numerical models with a common period of 30 years.

3.3. Defining the Multivariate Scenarios (Steps 3–5)

3.3.1. Dependency Modeling and Monte Carlo Simulations

First, the variables X_i for (Hs; SWL; U) are transformed onto the Gumbel scales using the standard probability integral transformation to give the variables Y_i. If Y_{-i} denotes a vector for all variables excluding Y_i, a multivariate non-linear regression model is applied to $Y_{-i} \mid Y_i > v$:

$$Y_{-i} = aY_i + Y_i^b W \quad \text{for } Y_i > v \text{ and } Y_i > Y_{-i} \tag{2}$$

where a and b are vectors of parameters, v a threshold, and W is a vector of residuals. Each fitted model describes the dependence between the variables (except Y_i). The models were adjusted using the maximum likelihood method and assuming that the residuals W are Gaussian. Using the diagnostic tools described by H&T04 [12], we here have selected $v = 0.95$ (expressed as a probability of non-exceeding this threshold). Examples of the parameters a and b, and of the selected thresholds v are given in Appendix D for the western and eastern region of Corsica.

Once the parameters of the dependency relationship were estimated, a 10,000-year period was simulated applying a Monte Carlo method. We obtained a set of fictive sea states, wind, and SWL conditions as illustrated in blue in Figure 4. The simulated variables Y_i are then transformed back to the original scales. Finally, the output is a large sample of sea and wind conditions (119,300 for the dataset NWW3 560/MARS 8) where at least one variable is extreme (i.e., exceeding a defined threshold) with respect to the individual marginal distributions, as well as the dependency relationship between variables.

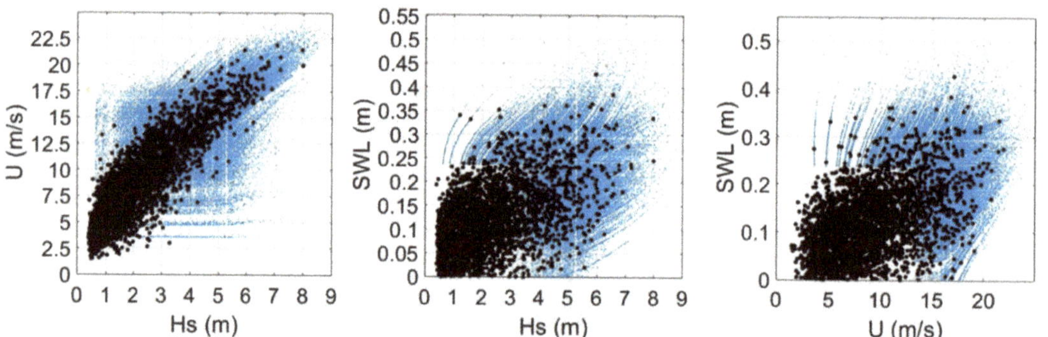

Figure 4. Illustration of Monte Carlo simulation for variables Hs, U, and SWL based on 30 common years of declustered data (black dots). Simulated data (10,000 years simulated) are in blue for the dataset NWW3 560/MARS 8.

3.3.2. Joint Probability Contours

Since we have our large sample obtained by Monte Carlo simulation, we can access the joint exceedance isocontours [22]. These 3D isocontours (x; y; z) are in the space of the three variables (Hs; SWL; U) where each point of the contour has the same probability of being exceeded:

$$P(\text{Hs} > \text{x, SWL} > \text{y, U} > \text{z}) = \frac{1}{\lambda T} \tag{3}$$

where λ is the number of events per year (in this case $\lambda = 122$) and T the return period considered in years. Figure 5a gives a 2D representation of the isocontours. The values on the isocontours are the SWL values from 0.2 to 0.4 m. From Figure 5a, it is then possible to identify combinations (Hs; SWL; U) of 100-year joint exceedance return periods. Figure 5b is a 3D representation of 100-year joint exceedance isocontours for the dataset NWW3 560/MARS 8. The black dots represent the combinations taken to force the hydrodynamic models. We have selected 34 scenarios for the dataset NWW3 560/MARS 8. These were selected along the isocontours by focusing preferably in the regions where the joint effect of the different drivers is the largest, i.e., where the convexity of the isocontours is the largest.

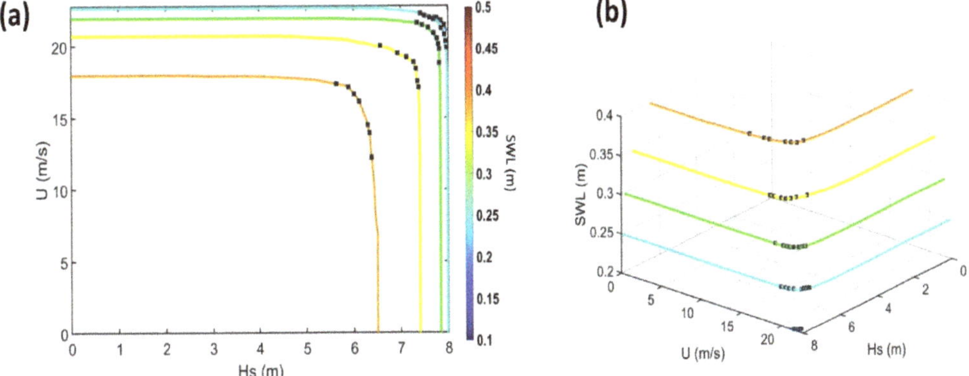

Figure 5. (**a**) Representation of the 100-year joint exceedance isocontours for the dataset NWW3 560/MARS 8. The color of the isocontours represent the SWL values from 0.2 to 0.4 m. (**b**) Simplified 3D representation of the 100-year joint exceedance isocontours for the dataset NWW3 560/MARS 8. The black dots represent the selected combinations given as input of the hydrodynamic models.

3.3.3. Covariates

Once the (Hs, SWL, U) combinations are identified each Hs must be associated to their peak period (Tp). The peak period (Tp) was simulated following the approach described by [37] and adapted from [22]. In this approach, a regression model of steepness (S_t), conditional on Hs is used:

$$S_t = \frac{2\pi\,Hs}{gT_p^2} \tag{4}$$

This model was used to simulate the steepness (and Tp) from Monte-Carlo simulations of Hs (blue dots in Figure 6). In order to associate Tp to Hs, we take values around the median of the simulated periods for each of the Hs considered. Table 2 gives the combinations Hs, Tp, U, and SWL selected from the NWW3 560/MARS 8 dataset, and representing the conditions offshore the Gulf of Porto.

For the remaining covariates, namely peak directions (Dp) associated to waves, and wind directions (Du), we have made the choice to impose the dominant direction(s) leading to extreme Hs and U. Figure 3b,d showed that Dp and Du associated to extreme waves and winds can be carried by one dominant sector (directions between 250 and 300°). Figure 7 illustrates wave (Figure 7b) and wind roses (Figure 7c) at points NWW3 512 and NWW3 544 (see locations in Figure 7a) giving the dominant peak directions (Dp), and wind directions (Du). These points give two dominant directions of extremes for Hs and U. The most penalizing values in terms of effect on the coast come from the west–south–west for the waves and south-west for the winds for the point NWW3 512. We forced the hydrodynamic models (SWASH-2DH and SWAN) with these directions. If there were two penalizing directions, regarding the exposure of the coastline, the directions were taken into account by running the hydrodynamic models twice depending on the sector of extreme Dp and Du. We did this double analysis for the point NWW3 544 as well and at every point presenting two penalizing directions.

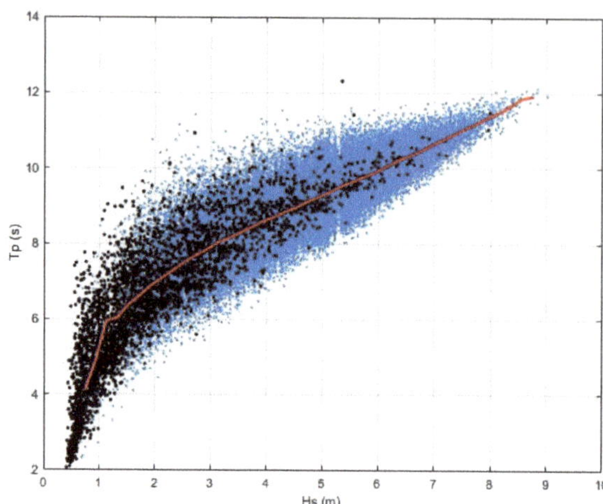

Figure 6. Relationship between Tp and Hs: declustered data (black dots), Monte Carlo simulation results (blue dots) for the dataset NWW3 560/MARS 8. The median is plotted in red.

Table 2. Combinations Hs, Tp, U, and SWL selected from the Gulf of Porto case.

Scenarios	Hs (m)	Tp (s)	U (m/s)	SWL (m)
1	7.85	11.22	21.76	0.2
2	7.76	11.09	21.95	0.2
3	7.92	11.39	21.45	0.2
4	7.7	11.22	21.85	0.25
5	7.61	10.79	21.98	0.25
6	7.51	10.91	22.1	0.25
7	7.42	10.71	22.25	0.25
8	7.86	11.53	21.25	0.25
9	7.93	11.27	20.81	0.25
10	7.95	11.32	20.35	0.25
11	7.97	11.3	19.89	0.25
13	7.59	11.04	21.27	0.3
14	7.46	11.14	21.44	0.3
15	7.35	11.05	21.61	0.3
16	7.7	10.9	20.9	0.3
17	7.77	11.48	20.52	0.3
18	7.79	11.18	20.13	0.3
19	7.82	11.31	19.75	0.3
20	7.82	11.15	18.79	0.3
21	6.94	10.58	19.46	0.35
22	6.57	10.39	20	0.35
23	7.13	11.01	19.19	0.35
24	7.28	10.59	18.85	0.35
25	7.33	11.07	18.4	0.35
26	7.36	10.77	17.52	0.35
27	7.38	11.13	17.1	0.35
28	6.01	9.55	16.63	0.4
29	5.89	9.65	17.12	0.4
30	5.64	9.91	17.35	0.4
31	6.11	10.41	16.12	0.4
32	6.29	10.08	14.49	0.4
33	6.33	9.91	13.94	0.4
34	6.37	9.9	12.24	0.4

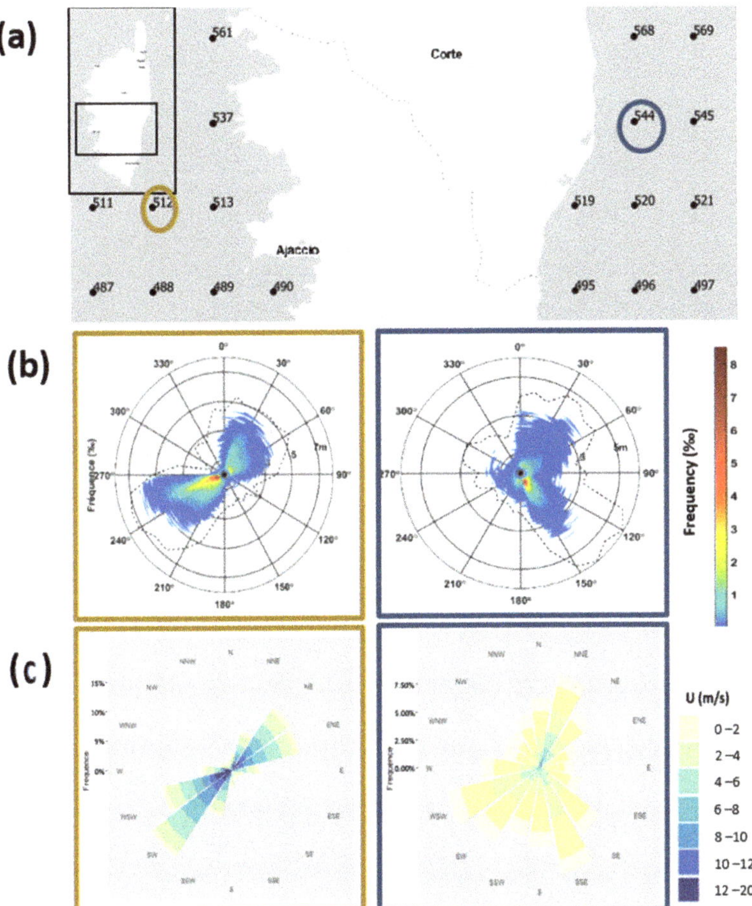

Figure 7. (a): Location of points NWW3 512 and NWW3 544 along Corsica. **(b)** Wave roses at point NWW3 512 (left) and NWW3 544 (right) giving the dominant peak direction (Dp) (°) and Hs frequency (‰). The black dashed lines represent the contour of extreme Hs. **(c)** Wind roses at point NWW3 512 (left) and NWW3 544 (right) giving the dominant wind direction Du (°), and U (m/s).

4. Results

Figure 8 shows the 100-year joint exceedance isocontours for all the points analyzed around Corsica. Different extreme regimes are here outlined: the dependence between the extreme triplets on the western part appear to be less marked than for the eastern part (as indicated by the convexity and range covered by the isocontours). This means that the 100-year triplets on the western coast deviate less from the 100-year return levels estimated using the marginal (i.e., without accounting for the dependence). This is clearly not the case on the eastern part as illustrated for instance by the Bastia analysis (black box) which shows large convexity of the isocontours. This 100-year joint exceedance isocontours' spatialization also highlights higher values of Hs (>6 m) and U (>18 m/s) offshore on the western part of the island compared to the eastern part of Corsica. This is in agreement with the different wind regimes (and induced waves) presented in Figure 1b. Indeed, on the west side of the island, wind and wave features (Hs > 6 m, U > 18 m/s, Tp > 12 s, 500 km < fetch < 1300 km) suggest that extreme scenarios may be driven by swells, whereas offshore conditions on the east part of Corsica mostly by wind waves and high still water levels. For example, on the eastern part of the island the combination (Hs = 2.2 m; Tp = 6.8 s; U = 12.1 s; SWL = 0.4 m), represented by a blue star in the red box in Figure 8, is identified as an extreme scenario to force the hydrodynamic models.

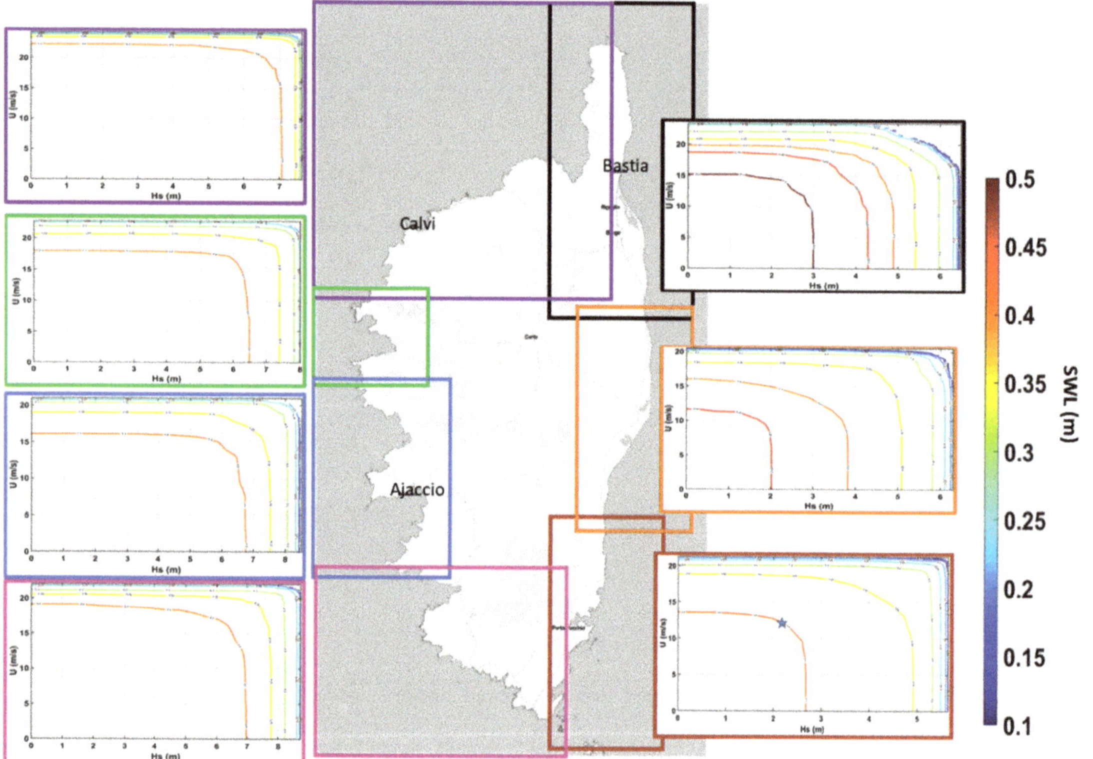

Figure 8. The 100-year joint exceedance isocontours for all of the boxes (in color) analyzed around Corsica, the colorbar corresponds to SWL (m).

The SWAN and SWASH-2DH simulations (steps 6–7) provide the contributions to compute total water levels (TWL) (step 8) at the shoreline for each offshore condition combination tested (see Section 2.2 for details on the methods). Figure 9 illustrates the distribution of TWL along the shoreline (a) for current conditions, and (b) for future conditions (2100). For current conditions, the static TWL along the shoreline in the study area is between 0.80 and 1.8 m. For "2100" future conditions, the values are between 1.2 and 2.2 m. Clearly, outside the area on the west of the Cap Corse in the north of the island, this value is hardly reached even under future conditions. Moreover, we note that values of TWL are higher on the west coast rather than the east coast, especially in the north west and south west, for both current and future conditions.

Finally, we tested the sensitivity of the results to the choices of the offshore conditions' combinations: this is generally below 0.05 m and can be considered not significant. Only a few combinations show differences greater than 0.05 m. In particular, Figure 10a shows that combinations 19 and 27 lead to the highest total water levels at the shoreline in the Gulf of Porto considering the specific directions Dp = 270° and Du = 240°. Moreover, the directions Dp and Du of waves and winds influence the scenarios leading to maximum water level at the shoreline. Indeed, when applying other directions (even with small changes) Dp = 240° and Du = 240° to force the SWAN and SWASH-2DH models, the scenario leading to the highest total water levels is different: combination 17 (Figure 10b). The differences are particularly marked in the southern Gulf of Porto where total levels at the shoreline are higher when applying Dp = 270° and Du = 240° (Figure 10c) instead of Dp = 240° and Du = 240° (Figure 10d).

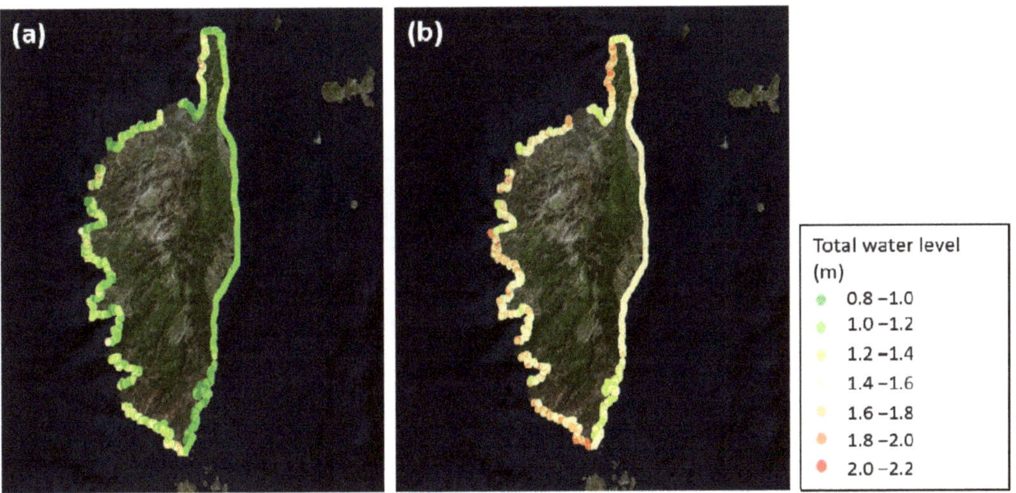

Figure 9. Total water level at the shoreline obtained: (**a**) for current conditions; (**b**) for future conditions (2100).

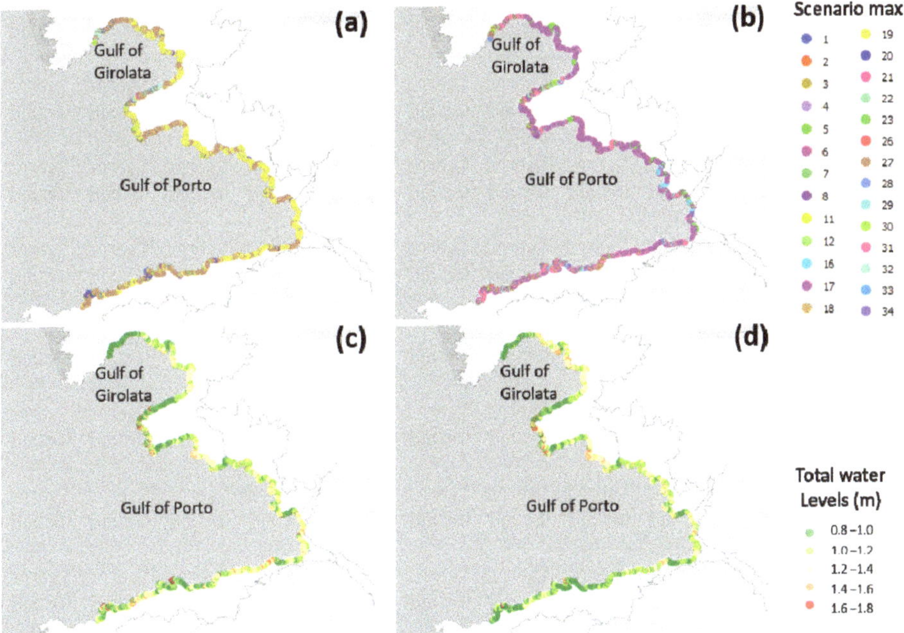

Figure 10. Scenarios leading to maximal water level at the shoreline in Gulf of Porto and Gulf of Girolata: (**a**) for directions Dp = 270° and Du = 240°; (**b**) for directions Dp = 240° and Du = 240° in current conditions. Total water level at the shoreline obtained in Gulf of Porto and Gulf of Girolata: (**c**) for directions Dp = 270° and Du = 240°; (**d**) for directions Dp = 240° and Du – 240° in current conditions.

5. Discussion and Conclusions

The key motivation of this study was to show how a multivariate extreme value analysis improve the estimate of low-lying zones potentially exposed to coastal flooding in Corsica. Indeed, in this zone, coastal sectors with altitudes below 2 m (and 2.4 m under "2100" future conditions), are considered potentially exposed to flooding. In order to asses if this value was relevant or not, we have first determined the joint probabilities that significant wave heights (Hs), wind intensity at 10 m above the ground (U), and water level (WL) exceed jointly imposed thresholds by applying a semiparametric multivariate extreme value analysis. Covariate peak direction (Dp), the peak period (Tp), and the wind direction (Du) were also accounted for. Once the (Hs, Tp, SWL, U) combinations were identified, we ran the coastal hydrodynamic models, SWAN and SWASH-2DH forced by these extreme scenarios. For current conditions, the results show that the TWL all along the shoreline in the study area are between 0.80 and 1.8 m, compared to 2 m currently applied on every low-lying zone. For "2100" future conditions, the values are between 1.2 and 2.2 m, compared to 2.4 m currently applied on every low-lying zone. In conclusion, the value 2 m (2.4 m when considering future conditions) seems to be overestimated regarding, the methods applied and the results of our study. This highlights the benefit of performing a full integration of extreme offshore conditions together with their dependence in hydrodynamic simulations for screening out the coastal areas potentially exposed to flooding around Corsica. More generally, the method can be adapted to areas where only topobathymetric studies have been carried out to map potential flooding areas.

There are however several aspects that might nuance this finding. First, uncertainty in the multivariate extreme value analysis was only partly integrated. Indeed, although we have taken into account some of the uncertainties (CI given with marginals, correction on Hs, and diagnosis on thresholds in H&T04), a full propagation of all the sources of uncertainty has not been performed. Thus, a simple-but-efficient approach consisting in affecting a 0.25 m margin on the TWL results showed that even in this situation, the 2 m threshold was hardly reached in several places around the island. Second, we did not apply the method in some areas like some cliffs or areas where topobathymetric data where insufficient at the time of the study. This constitutes a line for future research. Furthermore, the uncertainties also lie in the variables studied and the choice of dominant variable. Indeed, other variables could be studied, such as precipitation, river discharges, or even storm duration the same way as Hs [54]. A sensitivity analysis on these variables could be conducted on this area as in [55]. We can also note that we defined an independent wave event as a 3-day window (see Section 3.1), but [56] note that 25% of extreme wave events are consecutive events that hit the same location in less than a day. This could be relevant, to be taken into consideration, especially for "2100" future conditions (the 25% rate could be higher).

There are other points that we still need to work on. Indeed, we used the hydrodynamic models in a static way; it means that the propagation from offshore to nearshore does not take into account the dynamics of the phenomenon (dynamics of the event, flow velocity, etc.). With our static approach, the dynamic of an event is not taken into account and therefore the volume of water is only limited by the capacity of the low lying zone to fill up to an altitude corresponding to the total water level at the shoreline. Even if we attached importance to the dependency between several variables leading to flooding, the main component to determine low-lying zones potentially exposed to coastal flooding remains the topography. It could be relevant to consider a dynamic approach, which is more computationally demanding, but more precise in terms of overflow, breach flooding, volume of water assessment, or flood phenomena coming from rivers. Regarding methodological developments, future work could also concentrate on the comparison with alternatives approaches, such as a response approach [21], potentially aided by metamodeling techniques [57].

Furthermore, we looked at "2100" future conditions and, following the regulator's recommendations [46], we applied a 0.6 m value by 2100 to account the SLR in the hydrodynamic models. Even if this value was derived from the RCP 8.5 scenario "likely range", it is interesting to note that, regarding national SLR planning, France has a low amount of SLR used in planning compared to neighboring countries. Indeed, the amount of SLR used in planning in France is less than 1 m by 2100, whereas in Greece, it is about 1 m by 2100, and in Belgium, it is almost 2 m when considering 2100 high-end SLR [58]. Thiéblemont et al. showed that under the RCP 8.5, the 0.6 m value could be possibly doubled or tripled along the France coastline when considering high-end estimates of different components of sea-level projections [59]. Besides, the recent Special Report on the Ocean and Cryosphere in a Changing Climate (SROCC), then the Sixth Assessment Report AR6 emphasized that there is a substantial likelihood that SLR will be outside the likely range [60,61]. We note that these reports also identify deep uncertainties related to rare or indirectly deduced phenomenon, as the marine ice cliff instabilities taken into account in some global climate models.

Finally, the purpose here was the assessment of the total water level at the shoreline in order to inform on the low-lying zones in Corsica potentially exposed to flooding. The next steps of the method were to determine areas that were geomorphologically homogenous based on orthophotos and fieldwork analyses, in order to complete the mapping [62]. Specific information on the swash were also computed using the SWASH model in its 1D configuration on some beaches identified as potentially impacted by swash and mechanical shocks due to waves, in current and/or future conditions.

Author Contributions: Conceptualization, J.L., J.R., T.B. and R.P.; methodology, J.L., J.R., T.B., R.P., A.M. and J.M.; software, J.L., J.R., T.B., F.B.; validation, J.L.; formal analysis, J.L., J.R., T.B.; investigation, J.L., J.R., T.B., R.P., F.B., A.M. and J.M.; resources, J.L., J.R., T.B., F.B., R.P., A.M. and J.M.; data curation, J.L. and A.M.; writing—original draft preparation, J.L. and J.R.; writing—review and editing, J.L., J.R., T.B., R.P., F.B., A.M. and J.M.; project administration, J.M. All authors have read and agreed to the published version of the manuscript.

Funding: This research was co-funded by the DDTM Corse-du-Sud (AP16CSC008, AP17BAS013) and by the DREAL Haute-Corse and the BRGM (AP18BAS019).

Institutional Review Board Statement: Not applicable.

Informed Consent Statement: Not applicable.

Data Availability Statement: The data used is available on request from the providers mentioned.

Acknowledgments: The following data providers are acknowledged: Cerema (Candhis), CNES/AVISO, the NOAA, IGN, ISPRA, Météo-France, the Shom. The authors are very grateful to the anonymous reviewers for their comments and expert advice. We also thank F. Paris for his support on the preliminary phase of this study.

Conflicts of Interest: The authors declare no conflict of interest. The funders had no role in the design of the study; in the collection, analyses, or interpretation of data; in the writing of the manuscript, or in the decision to publish the results.

Appendix A. Validation of the Numerical Models

To validate the hydrodynamic model, we compared still water level observations at the Ajaccio tide gauge to the simulations.

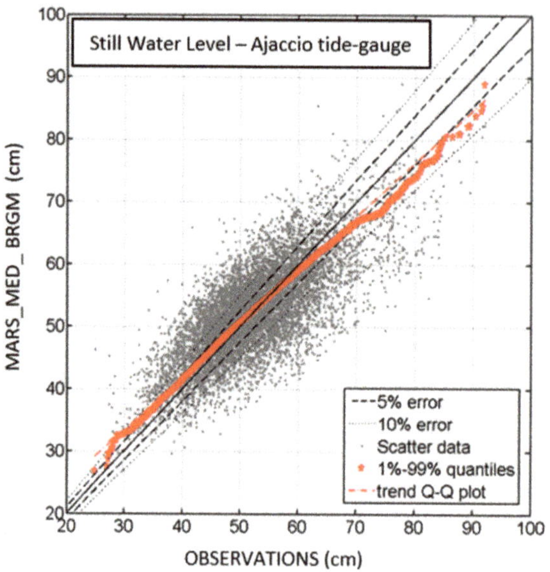

Figure A1. Q–Q plot and scatter plot of hourly still water levels measured at Ajaccio tide gauge (OBSERVATIONS) and values obtained by numerical simulation (MARS_MED_BRGM).

A linear correction is applied to the NWW3 MED data. This correction is derived from the linear regression between the observations and the original model data and is expressed as follows:

$$\text{NWW3 MED corrected} = 1.21 \times \text{NWW3 MED} + 0.14 \tag{A1}$$

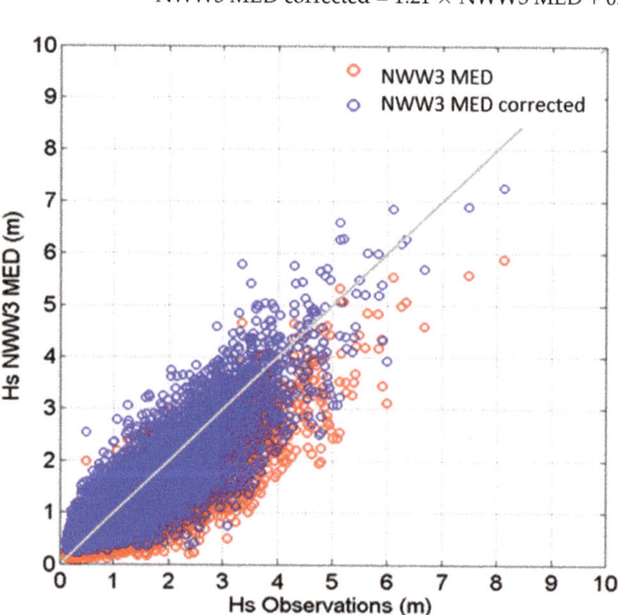

Figure A2. Observed and simulated Hs at the Cap Corse buoy, period 1999–2008.

Appendix B. Elements Used for Adjusting Marginal Probability Distributions for Hs, SWL, and U around Corsica

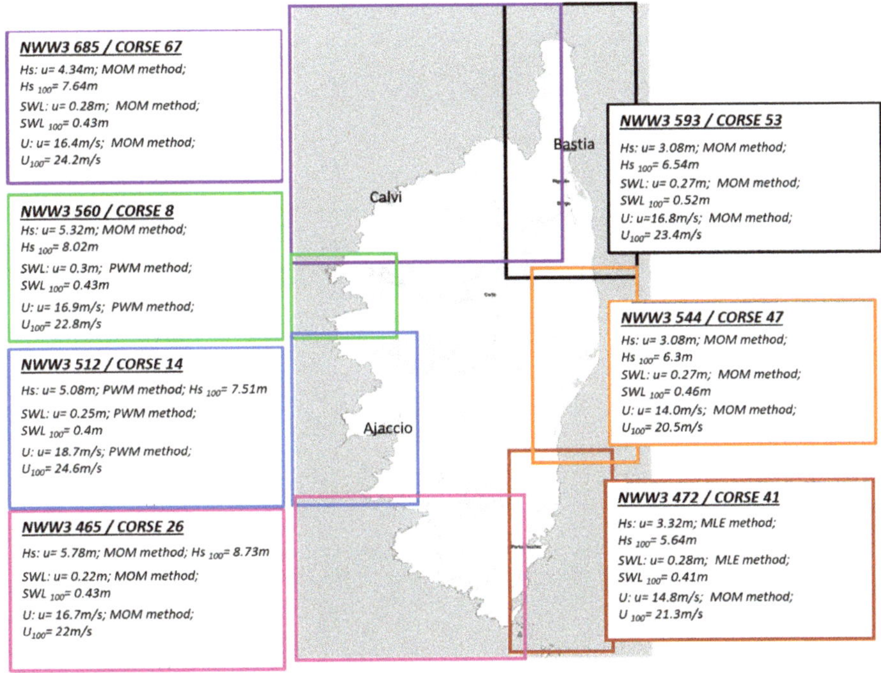

Figure A3. Selected thresholds u, methods used (MOM, PWM, MLE) to estimate GPD parameters, and the resulting 100-year return level for Hs (m), SWL (m), and U (m/s).

Appendix C. Wave Directions Affecting the Coastline for Extreme Conditions

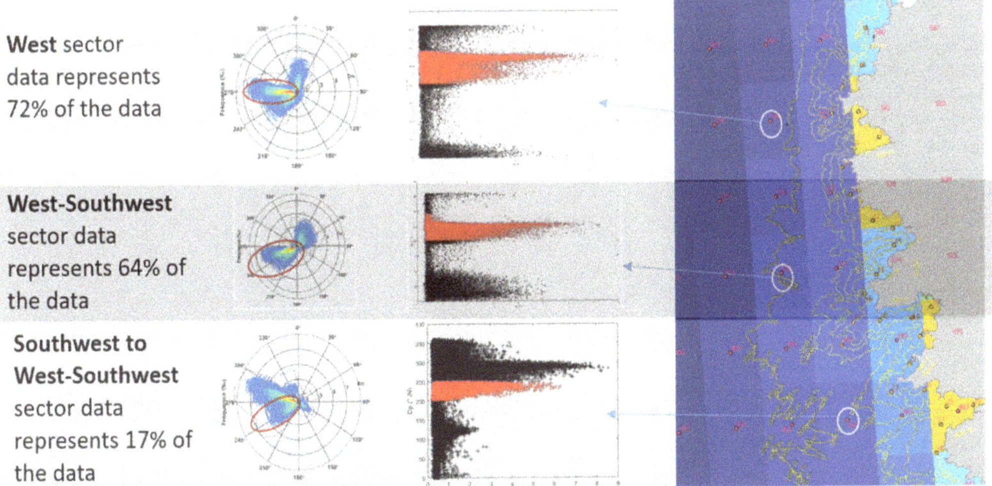

Figure A4. Wave roses; and Dp(°) plotted against Hs(m) for three offshore points. The red dots and red circles indicate the wave directions, which could most affect the coastline for extreme conditions regarding the coast morphology.

Appendix D. Dependence Coefficients

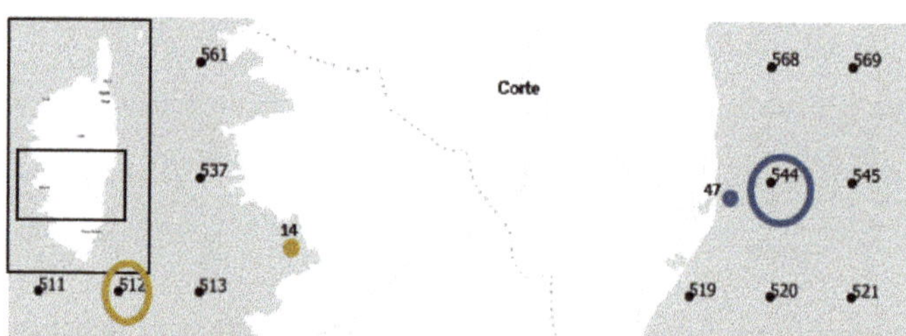

	NWW3 512 / CORSE 14: v=0,99			NWW3 544 / CORSE 47: v=0,96		
Pairs	**a coefficient**	**b coefficient**	**W residuals**	**a coefficient**	**b coefficient**	**W residuals**
U|Hs	0.83	1	0.37	0.52	0.58	0.55
SWL|Hs	0.5	1	0.12	0.4	0.03	0.23
Hs|U	0.84	1	0.65	0.84	1	0.94
SWL|U	0.33	0.03	0.32	0.32	0.28	0.77
Hs|SWL	0.52	0.85	0.64	0.07	0.55	0.84
U|SWL	0.42	1	0.39	0.21	0.13	0.09

Figure A5. Dependence coefficients a and b, residuals Z for selected threshold ν for two coastal regions of Corsica.

References

1. Cagigal, L.; Rueda, A.; Anderson, D.; Ruggiero, P.; Merrifield, M.A.; Montaño, J.; Coco, G.; Méndez, F.J. A multivariate, stochastic, climate-based wave emulator for shoreline change modelling. *Ocean. Model.* **2020**, *154*, 101695. [CrossRef]
2. Marcos, M.; Rohmer, J.; Vousdoukas, M.I.; Mentaschi, L.; Le Cozannet, G.; Amores, A. Increased Extreme Coastal Water Levels Due to the Combined Action of Storm Surges and Wind Waves. *Geophys. Res. Lett.* **2019**, *46*, 4356–4364. [CrossRef]
3. Galiatsatou, P.; Prinos, P. Estimation of extreme storm surges using a spatial linkage assumption. In Proceedings of the 32nd IAHR Congress, Venice, Italy, 1–6 July 2007.
4. Zachary, S.; Feld, G.; Ward, G.; Wolfram, J. Multivariate extrapolation in the offshore environment. *Appl. Ocean. Res.* **1998**, *20*, 273–295. [CrossRef]
5. Gouldby, B.; Wyncoll, D.; Panzeri, M.; Franklin, M.; Hunt, T.; Hames, D.; Tozer, N.; Hawkes, P.; Dornbusch, U.; Pullen, T. Multivariate extreme value modelling of sea conditions around the coast of England. *Proc. Inst. Civ. Eng. Marit. Eng.* **2017**, *170*, 3–20. [CrossRef]
6. Corbella, S.; Stretch, D.D. Simulating a multivariate sea storm using Archimedean copulas. *Coast. Eng.* **2013**, *76*, 68–78. [CrossRef]
7. Wahl, T.; Mudersbach, C.; Jensen, J. Assessing the hydrodynamic boundary condi-tions for risk analyses in coastal areas: A multivariate statistical approach based oncopula functions. *Nat. Hazards Earth Syst. Sci.* **2012**, *12*, 495–510. [CrossRef]
8. De Michele, C.; Salvadori, G.; Passoni, G.; Vezzoli, R. A multivariate model of sea storms using copulas. *Coast. Eng.* **2007**, *54*, 734–751. [CrossRef]
9. Soares, C.G.; Cunha, C. Bivariate autoregressive models for the time series of significant wave height and mean period. *Coast. Eng.* **2000**, *40*, 297–311. [CrossRef]
10. Jonathan, P.; Ewans, K. Statistical modelling of extreme ocean environments for marine design: A review. *Ocean. Eng.* **2013**, *62*, 91–109. [CrossRef]
11. Idier, D.; Rohmer, J.; Pedreros, R.; Le Roy, S.; Lambert, J.; Louisor, J.; Le Cozannet, G.; Le Cornec, E. Coastal Flood: A Composite Method for Past Events Characterisation Providing Insights in Past, Present and Future Hazards—Joining Historical, Statistical and Modelling Approaches. *Nat. Hazards* **2020**, *101*, 465–501. [CrossRef]
12. Heffernan, J.E.; Tawn, J.A. A conditional approach for multivariate extreme values (with discussion). *J. R. Stat. Soc. Ser. B Statist. Methodol.* **2004**, *66*, 497–546. [CrossRef]
13. Monbet, V.; Ailliot, P.; Prevosto, M. Survey of stochastic mod-els for wind and sea state time series. *Prob. Eng. Mech.* **2007**, *22*, 113–126. [CrossRef]
14. Ewans, K.; Jonathan, P. Evaluating environmental joint extremes for the offshore industry using the conditional extremes model. *J. Mar. Syst.* **2013**, *130*, 124–130. [CrossRef]
15. Jonathan, P.; Ewans, K.; Randell, D. Joint modelling of extreme ocean environments incorporating covariate effects. *Coast. Eng.* **2013**, *79*, 22–31. [CrossRef]
16. Jonathan, P.; Flynn, J.; Ewans, K. Joint modelling of wave spectral parameters for extreme sea states. *Ocean. Eng.* **2013**, *37*, 1070–1080. [CrossRef]

17. Stepanian, A.; Balouin, Y.; Belon, R.; Bodéré, G. ROL—Etude Complémentaire sur le Littoral de la Plaine Orientale de Corse—Etat des Connaissances sur les Impacts des Tempêtes sur le Littoral. Rapport Final; Rapport BRGM/RP-59046-FR. 2011, p. 137. Available online: http://infoterre.brgm.fr/rapports/RP-59046-FR.pdf (accessed on 23 July 2018).

18. Western European Union. Wind and Wave Atlas of the Mediterranean Sea. *Technical Note* 2004. Available online: http://users.ntua.gr/mathan/pdf/Pages_from%20_WIND_WAVE_ATLAS_MEDITERRANEAN_SEA_2004.pdf. (accessed on 21 June 2018).

19. Krestenitis, Y.; Pytharoulis, I.; Karacostas, T.S.; Androulidakis, Y.; Makris, C.; Kombiadou, K.; Tegoulias, I.; Baltikas, V.; Kotsopoulos, S.; Kartsios, S. *Severe Weather Events and Sea Level Variability Over the Mediterranean Sea: The WaveForUs Operational Platform*; Springer: Berlin/Heidelberg, Germany, 2017; pp. 63–68. [CrossRef]

20. Mentaschi, L.; Besio, G.; Cassola, F.; Mazzino, A. Performance evaluation of Wavewatch III in the Mediterranean Sea. *Ocean. Model.* **2015**, *90*, 82–94. [CrossRef]

21. Sanuy, M.; Jiménez, J.A.; Ortego, M.I.; Toimil, A. Differences in assigning probabilities to coastal inundation hazard estimators: Event versus response approaches. *J. Flood Risk Manag.* **2020**, *13*. [CrossRef]

22. Hawkes, P.J.; Gouldby, B.; Tawn, J.A.; Owen, M.W. The joint probability of waves and water levels in coastal engineering design. *J. Hydraul. Res.* **2002**, *40*, 241–251. [CrossRef]

23. Nicolae Lerma, A.; Bulteau, T.; Elineau, S.; Paris, F.; Durand, P.; Anselme, B.; Pedreros, R. High-resolution marine flood modelling coupling overflow and overtopping processes: Framing the hazard based on historical and statistical ap-proaches. *Nat. Haz. Earth Syst. Sci.* **2018**, *18*, 207–229. [CrossRef]

24. Pillet, D.; Renoult, R.; Saillard, M. Audit de Suivi de la Mise en Oeuvre de la Politique de Prévention des Risques Naturels et Technologiques dans la Région Corse. Institutional Report. 2019, pp. 12–13. Available online: https://www.vie-publique.fr/sites/default/files/rapport/pdf/194000570.pdf (accessed on 11 February 2021).

25. Shom—Ouvrage de Marée. References Altimétriques Maritimes Ports de France Métropolitaine et d'Outre-Mer. *Cotes du Zéro Hydrographique et Niveaux Caractéristiques de la Marée*, 2017th ed.; 2017; p. 120. Available online: https://diffusion.shom.fr/pro/references-altimetriques-maritimes-ram.html (accessed on 27 March 2018).

26. Haigh, I.D.; Nicholls, R.; Wells, N. A comparison of the main methods for estimating probabilities of extreme still water levels. *Coast. Eng.* **2010**, *57*, 838–849. [CrossRef]

27. Kergadallan. *Analyse Statistique des Niveaux d'Eau Extrêmes—Environnements Maritime et Estuarien*; Technical Note; CETMEF: Paris, France, 2013.

28. Chawla, A.; Spindler, D.M.; Tolman, H.L. *WAVEWATCH III Hindcasts with Reanalysis Winds. Initial Report on Model Setup*; Technical Note 291, NOAA/NWS/NCEP/MMAB; U.S. Department of Commerce, National Oceanic and Atmospheric Administration: Camp Springs, MD, USA, 2011; p. 100.

29. Chawla, A.; Spindler, D.M.; Tolman, H.L. *30 Year Wave Hindcasts Using WAVEWATCH III R with CFSR Winds, Phase 1*; Technical Note 302; NOAA/NWS/NCEP/MMAB; U.S. Department of Commerce, National Oceanic and Atmospheric Administration: Camp Springs, MD, USA, 2012; p. 23.

30. Ardhuin, F.; Rogers, A.E.; Babanin, J.-F.; Filipot, R.; Magne, A.; Roland, A.; van der Westhuysen, P.; Queffeulou, J.-M.; Lefevre, L.; Aouf, F.; et al. Semi-empirical dissipation source functions for wind-wave models: Part I, definition, calibration and validation. *J. Phys. Oceanogr.* **2010**, *40*, 1917–1941. [CrossRef]

31. Lazure, P.; Dumas, F. An external–internal mode coupling for a 3D hydrodynamical model for applications at regional scale (MARS). *Adv. Water Res.* **2008**, *31*, 233–250. [CrossRef]

32. Saha, S.; Moorthi, S.; Pan, H.; Wu, X.; Wang, J.; Nadiga, S.; Tripp, P.; Kistler, R.; Woollen, J.; Behringer, D.; et al. *NCEP Climate Forecast. System Reanalysis (CFSR) Selected Hourly Time-Series Products, January 1979 to December 2010*; Research Data Archive at the National Center for Atmospheric Research, Computational and Information Systems Laboratory: Boulder, CO, USA, 2010. [CrossRef]

33. Shom (2015)—MNT Bathymétrique de Façade de la Corse (Projet Homonim). Available online: https://diffusion.shom.fr/pro/risques/bathymetrie/mnt-facade-atl-homonim-264.html (accessed on 14 May 2019).

34. Brodtkorb, P.; Johannesson, A.; Lindgren, P.; Rychlik, G.; Rydén, I.; Sjö, J.; Sjö, E. WAFO—A Matlab Toolbox for Analysis of Random Waves And Loads. In Proceedings of the Tenth International Offshore and Polar Engineering Conference, Seattle, WA, USA, 28 May–2 June 2000.

35. Guanche Garcia, Y.; Prevosto, M.; Maisondieu, C.; Bulteau, T.; Paris, F. *Analyses of Environmental Time Series: Extreme Values*; Scientific Report; Ifremer: Brest, France, 2015. [CrossRef]

36. Efron, B. Bootstrap Methods: Another Look at the Jackknife. In *Springer Series in Statistics*; Springer: Berlin/Heidelberg, Germany, 1992; pp. 569–593.

37. Wyncoll, D.; Gouldby, B. Integrating a multivariate extreme value method within a system flood risk analysis model. *J. Flood Risk Manag.* **2015**, *8*, 145–160. [CrossRef]

38. Gouldby, B.; Mendez, F.; Guanche, Y.; Rueda, A.; Mínguez, R. A methodology for deriving extreme nearshore sea conditions for structural design and flood risk analysis. *Coast. Eng.* **2014**, *88*, 15–26. [CrossRef]

39. Booij, N.; Ris, R.C.; Holthuijsen, L.H. A third-generation wave model for coastal regions, Part I: Model description and validation. *J. Geophys. Res.* **1999**, *104*, 7649–7666. [CrossRef]

40. Smit, P.; Zijlema, M.; Stelling, G. Depth-induced wave breaking in a non-hydrostatic, near-shore wave model. *Coast. Eng.* **2013**, *76*, 1–16. [CrossRef]

41. Stelling, G.; Zijlema, M. An accurate and efficient finite-difference algorithm for non-hydrostatic free-surface flow with application to wave propagation. *Int. J. Num. Methods Fluids* **2003**, *43*, 1–23. [CrossRef]

42. Stelling, G.S.; Duinmeijer, S.P.A. A staggered conservative scheme for every Froude number in rapidly varied shallow water flows. *Int. J. Num. Methods Fluids* **2003**, *43*, 1329–1354. [CrossRef]

43. Zijlema, M.; Stelling, G.S. Further experiences with computing non-hydrostatic free-surface flows involving water waves. *Int. J. Num. Methods Fluids* **2005**, *48*, 169–197. [CrossRef]

44. Zijlema, M.; Stelling, G. Efficient computation of surf zone waves using the nonlinear shallow water equations with non-hydrostatic pressure. *Coast. Eng.* **2008**, *55*, 780–790. [CrossRef]

45. Zijlema, M.; Stelling, G.; Smit, P. SWASH: An operational public domain code for simulating wave fields and rapidly varied flows in coastal waters. *Coast. Eng.* **2011**, *58*, 992–1012. [CrossRef]

46. MEDDE Guide Méthodologique: Plan de Prévention des Risques Littoraux. 2014, p. 169. Available online: https://www.ecologie.gouv.fr/sites/default/files/Guide%20PPRL%20-%20version%20finale%20mai%202014.pdf (accessed on 22 December 2020).

47. Church, J.A.; Clark, P.U.; Cazenave, A.; Gregory, J.M.; Jevrejeva, S.; Levermann, A.; Merrifield, M.A.; Milne, G.A.; Nerem, R.S.; Nunn, P.D.; et al. Sea Level Change. In *Climate Change 2013: The Physical Science Basis. Contribution of Working Group I to the Fifth Assessment Report of the Intergovernmental Panel on Climate Change*; Stocker, T.F., Qin, D., Plattner, G.-K., Tignor, M., Allen, S.K., Boschung, J., Nauels, A., Xia, Y., Bex, V., Midgley, P.M., Eds.; Cambridge University Press: Cambridge, UK; New York, NY, USA, 2013.

48. Tebaldi, C.; Kopp, R.E.; Horton, R.M.; Little, C.M.; Mitrovica, J.X.; Oppenheimer, M.; Rasmussen, D.J.; Strauss, B.H. Probabilistic 21st and 22nd century sea-level projections at a global network of tide-gauge sites. *Adv. Earth Space Sci.* **2014**, *2*, 383–406. [CrossRef]

49. Slangen, A.B.A.; Carson, M.; Katsman, C.A.; van de Wal, R.S.W.; Köhl, A.; Vermeersen, L.L.A.; Stammer, D. Projecting twenty-first century regional sea-level changes. *Clim. Chang.* **2014**, *124*, 317–332. [CrossRef]

50. Coles, S.; Bawa, J.; Trenner, L.; Dorazio, P. *An Introduction to Statistical Modeling of Extreme Values*; Springer: London, UK, 2001; Volume 208.

51. Ullmann, A.; Pirazzoli, P.; Moron, V. Sea surges around the Gulf of Lions and atmospheric conditions. *Glob. Planet. Chang.* **2008**, *63*, 203–214. [CrossRef]

52. Hosking, J.R.M.; Wallis, J.R. Parameter and quantile estimation for the generalized Pareto distribution. *Technometrics* **1987**, *29*, 339–349. [CrossRef]

53. Mazas, F.; Hamm, L. A multi-distribution approach to POT methods for determining extreme wave heights. *Coast. Eng.* **2011**, *58*, 385–394. [CrossRef]

54. Callaghan, D.; Nielsen, P.; Short, A.; Ranasinghe, R. Statistical simulation of wave climate and extreme beach erosion. *Coast. Eng.* **2008**, *55*, 375–390. [CrossRef]

55. Camus, P.; Haigh, I.D.; Nasr, A.A.; Wahl, T.; Darby, S.E.; Nicholls, R.J. Regional analysis of multivariate compound coastal flooding potential around Europe and environs: Sensitivity analysis and spatial patterns. *Nat. Hazards Earth Syst. Sci.* **2021**, *21*, 2021–2040. [CrossRef]

56. Martzikos, N.T.; Prinos, P.E.; Memos, C.D.; Tsoukala, V.K. Statistical analysis of Mediterranean coastal storms. *Oceanologia* **2021**, *63*, 133–148. [CrossRef]

57. Rohmer, J.; Idier, D. A meta-modelling strategy to identify the critical offshore conditions for coastal flooding. *Nat. Hazards Earth Syst. Sci.* **2012**, *12*, 2943–2955. [CrossRef]

58. McEvoy, S.; Haasnoot, M.; Biesbroek, R. How are European countries planning for sea level rise? *Ocean. Coast. Manag.* **2021**, *203*, 105512. [CrossRef]

59. Thiéblemont, R.; Le Cozannet, G.; Toimil, A.; Meyssignac, B.; Losada, I.J. Likely and High-End Impacts of Regional Sea-Level Rise on the Shoreline Change of European Sandy Coasts Under a High Greenhouse Gas Emissions Scenario. *Water* **2019**, *11*, 2607. [CrossRef]

60. Oppenheimer, M.; Glavovic, B.C.; Hinkel, J.; van de Wal, R.; Magnan, A.K.; Abd-Elgawad, A.; Cai, R.; CifuentesJara, M.; DeConto, R.M.; Ghosh, T.; et al. Sea Level Rise and Implications for Low-Lying Islands, Coasts and Communities. In *IPCC Special Report on the Ocean and Cryosphere in a Changing Climate*; Pörtner, H.-O., Roberts, D.C., Masson-Delmotte, V., Zhai, P., Tignor, M., Poloczanska, E., Mintenbeck, K., Alegría, A., Nicolai, M., Okem, A., et al., Eds.; Intergovernmental Panel on Climate Change: Geneva, Switzerland, 2019.

61. IPCC. *Climate Change 2021: The Physical Science Basis. Contribution of Working Group I to the Sixth Assessment Report of the Intergovernmental Panel on Climate Change*; Masson-Delmotte, V., Zhai, P., Pirani, A., Connors, S.L., Péan, C., Berger, S., Caud, N., Chen, Y., Goldfarb, L., Gomis, M.I., et al., Eds.; Cambridge University Press: Cambridge, UK, 2021.

62. Tonisson, H.; Suursaar, U.; Kont, A. Maps, aerial photographs, orthophotos and GPS data as a source of information to determine shoreline changes, coastal geomorphic processes and their relation to hydrodynamic conditions on Osmussaar Island, the Baltic Sea. In Proceedings of the 2012 IEEE International Geoscience and Remote Sensing Symposium, Munich, Germany, 22–27 July 2012; pp. 2657–2660. [CrossRef]

Journal of
*Marine Science
and Engineering*

Article

On-Site Investigations of Coastal Erosion and Accretion for the Northeast of Taiwan

Ting-Yu Liang [1], Chih-Hsin Chang [1], Shih-Chun Hsiao [2], Wei-Po Huang [3], Tzu-Yin Chang [1], Wen-Dar Guo [1], Che-Hsin Liu [1], Jui-Yi Ho [1] and Wei-Bo Chen [1,*]

[1] National Science and Technology Center for Disaster Reduction, New Taipei City 23143, Taiwan; lty@ncdr.nat.gov.tw (T.-Y.L.); chang.c.h@ncdr.nat.gov.tw (C.-H.C.); geoct@ncdr.nat.gov.tw (T.-Y.C.); wdguo@ncdr.nat.gov.tw (W.-D.G.); a120160@ncdr.nat.gov.tw (C.-H.L.); juiyiho@ncdr.nat.gov.tw (J.-Y.H.)
[2] Department of Hydraulic and Ocean Engineering, National Cheng Kung University, Tainan City 70101, Taiwan; schsiao@mail.ncku.edu.tw
[3] Department of Harbor and River Engineering, National Taiwan Ocean University, Keelung City 202301, Taiwan; a0301@mail.ntou.edu.tw
* Correspondence: wbchen@ncdr.nat.gov.tw; Tel.: +886-2-8195-8611

Abstract: Coastal erosion is a major natural hazard along the northeastern shoreline (i.e., Yilan County) of Taiwan. Analyses of the evolution of the 0 m isobath of the Yilan County coastline indicate that erosion and accretion are occurring north and south of Wushi Fishery Port, respectively, because of jetty and groin construction. Topographic and bathymetric surveys involving the measurement of 43 cross sections were conducted in 2006, 2012, 2013, and 2019. The cross-shore profile comparisons reveal that the erosion of onshore dunes is significant in the northern Jhuan River estuary. Due to the establishment of a nature reserve in the southern Lanyang River estuary, the sediments are carried northward by tidal currents, and accretion is inevitable in the northern Lanyang River estuary. The results of the bathymetric surveys also suggest that the shoreline of Yilan County tends to accrete in summer because of abundant sediment from the rivers; however, it is eroded in winter, owing to the large waves induced by the northeast monsoon. Additionally, the calculated net volume of erosion and accretion between each pair of cross sections shows that the length of coastline impacted by estuarine sediment transport is approximately 2 km long from north to south along the coastline of the Lanyang River estuary.

Keywords: erosion and accretion; cross-shore profile evolution; Lanyang River estuary; limit of estuarine sediment transport; northeastern coastal waters of Taiwan

Citation: Liang, T.-Y.; Chang, C.-H.; Hsiao, S.-C.; Huang, W.-P.; Chang, T.-Y.; Guo, W.-D.; Liu, C.-H.; Ho, J.-Y.; Chen, W.-B. On-Site Investigations of Coastal Erosion and Accretion for the Northeast of Taiwan. *J. Mar. Sci. Eng.* **2022**, *10*, 282. https://doi.org/10.3390/jmse10020282

Academic Editor: Rodger Tomlinson

Received: 7 February 2022
Accepted: 10 February 2022
Published: 18 February 2022

Publisher's Note: MDPI stays neutral with regard to jurisdictional claims in published maps and institutional affiliations.

1. Introduction

Coastal erosion can lead to coastal retreat, habitat destruction, and loss of land, which result in significant negative ecological and socioeconomic impacts on global coastal zones. Beach and dune systems are the first line for defending against the damaging impacts of water-related natural hazards, such as coastal storms, hurricanes, and typhoons; therefore, shoreline erosion (retreat) poses a significant threat to settled coastal areas worldwide [1,2]. Rapidly changing coastlines are a serious problem in many areas of the world, such as sections of the Nile and the Yellow River Delta, which have stirred the interest of many researchers in different fields [3–20].

The sediment budget and geology determine coastal morphology and dynamics, which influence the nature and health of coastal ecosystems. Human activities affecting sediment dynamics, both on the coast and on land, modify the naturally occurring patterns of erosion and accretion. Additionally, human interventions have frequently altered the delivery of riverine sediments to coastal areas [21–24]. For example, reservoir/dam construction has trapped over 50% of the world's sediment flux [25], and most of the world's deltas have now been significantly dammed in their upper and central reaches [21,25].

79

According to the report from Warrick et al. [26], approximately 30 million tons (Mt) of sediment was deposited in the reservoirs of the river they studied before dam removal began in 2011.

Many areas of observed historical shoreline advances are related to reclamation and impoundment by coastal structures. These human activities modify coastal dynamics, typically resulting in downdrift erosion. Factors that influence coastal erosion and sedimentation encompass characteristics of coastal sediment; exchanges among the land, the coast, and the shelf; geomorphic responses to oceanic forcing. Human activities may both substantially influence and be affected by coastal erosion and sedimentation [2,27,28]. Currently, climate change impacts, including sea-level rise and potential increases in the frequency and intensity of severe tropical and extratropical storms, hurricanes, and typhoons, could accelerate coastal erosion or accretion, and recent observations have also indicated an acceleration in coastal cliff erosion [29].

Dadson et al. [30] reported erosion rates in the Taiwan Mountains estimated from modern river sediment loads. They suggested that Taiwan supplied 384 Mt yr^{-1} of suspended sediment to the ocean from 1970 to 1999, which represents 1.9% of the estimated global suspended sediment discharge but is derived from only 0.024% of Earth's subaerial surface. A better understanding of the nature and evolution of coastal (beach) erosion and accretion is necessary to inform and enact appropriate and timely disaster preparedness [31,32]. Moreover, to accurately assess coastal hazards in the face of future climate and land-use changes, it is necessary to understand the dynamics of shoreline erosion and accretion over the length and time scales relevant to the processes that drive change.

Direct measurements through traditional on-boat acoustic surveys have a high resolution, although they consume considerable manpower and material resources [33]. The present study investigated the timing of the transition from either stable or erosional conditions to accretional conditions in the study area by evaluating a time series of cross-shore positions. The aim and insights derived from the present study are expected to clarify coastal, beach, estuarine, and tidal flat management strategies. The findings are also helpful for the development of similar coastal systems worldwide. The details of the study site information, the evolution of the 0 m isobath, on-site topography, and bathymetric survey are described in Section 2; the analyses of interannual and seasonal variability derived from the surveyed data are presented in Section 3. In Section 4, a discussion on the alongshore limit of the sediment delivered from the Lanyang River is given, and finally, the summary and conclusions are presented in Section 5.

2. Materials and Methods

2.1. Description of Study Site

Yilan County covers the entire shoreline of northeastern Taiwan, with a total length of 101 km. The whole coastline is categorized as a fault coast in northern and southern Yilan and an alluvial-plain coast in middle Yilan according to the geological features. The section of shoreline selected for topographic and bathymetric surveys extended from north of the Wushi Fishery Port to south of the Lanyang River estuary, with a length of nearly 20 km (red line in Figure 1a), and extended from the shoreline to approximately 800 to 1000 m offshore. Figure 1b shows an aerial image of the Wushi Fishery Port taken in the present study. As shown in Figure 1b, an offshore jetty and an extended jetty, with lengths of 500 m and 400 m, respectively, lie to the east and north of the Wushi Fishery Port, respectively, to stabilize the oscillation caused by waves. A wave buoy located west of Gueishan Island (the orange triangle in Figure 1a) is managed by the Central Weather Bureau (CWB) of Taiwan. The hourly measurements of significant wave height and current recorded at the buoy from 2006 to 2020 are graphed in Figure 1c,d, respectively. It is obvious that the predominant waves come from the northeast with a height of 1–2 m (Figure 1c); however, this coastal area is threatened with big waves (significant wave height > 3.0 m) during the passage of typhoons [34,35]. The principal currents range from 0.25 to 0.5 m/s, with a flow direction to the southeast (Figure 1d). Erosion and accretion phenomena are expected to be found in the north and south of Wushi Fishery Port, respectively, due to the physical characteristics of the marine environment and jetty effect. According to the report from the Water Resource Agency (WRA) of Taiwan, the Lanyang River is the largest river (with a watershed area of 978 km^2) in Yilan County and has an average annual runoff of 74.24 ms^{-1} and a sediment yield of 6 million m^3. An aerial image (taken by the present study) of the Lanyang River estuary is shown in Figure 1e, and a sandbar in the south of the Lanyang River estuary is clearly shown. Additionally, the daily sediment emissions from the Lanyang River estuary were measured by the WRA on the specified days in 2020. Figure 2 illustrates the daily variations of estuarine sediment volume at the Lanyang River. It can be seen that the months with higher daily sediment volume in the Lanyang River estuary are the end of April to September each year, due to their being wet and typhoon seasons. The median particle diameter (D50) of the Lanyang River estuary is 5.949 mm, according to the report from the WRA.

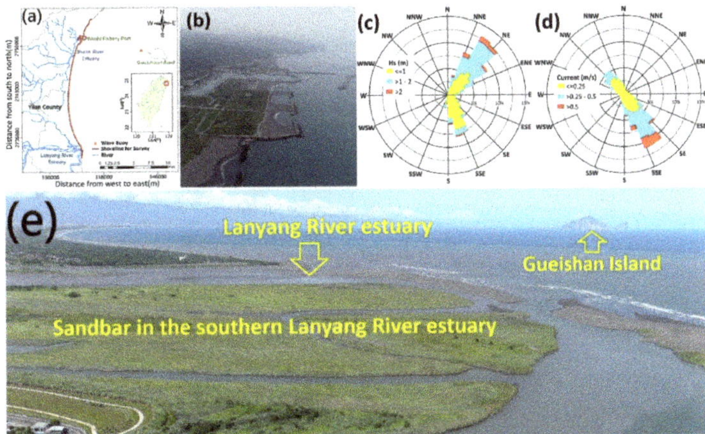

Figure 1. (**a**) The location of the study area, (**b**) an aerial image of the Wushi Fishery Port, (**c**) directional distribution of significant wave height, (**d**) directional distribution of current measured at a wave buoy near Guishan Island from 2006 to 2020, and (**e**) an aerial image of the Lanyang River estuary.

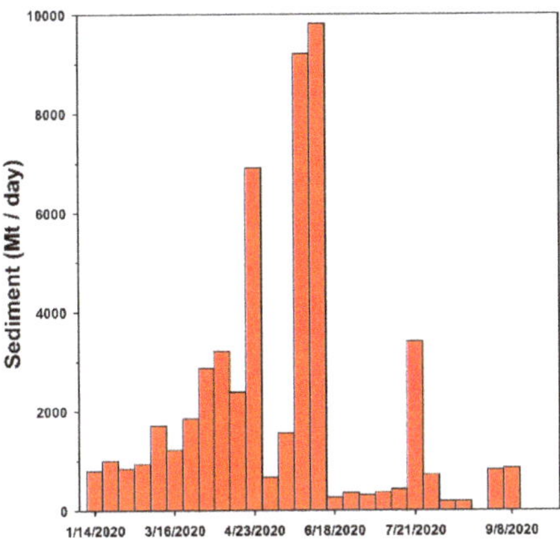

Figure 2. Measurements of daily sediment discharged from the Lanyang River estuary in 2020.

2.2. Evolution of the 0 m Isobath

To better understand the shoreline evolution in the coastal area with significant erosion and accretion phenomena, the 0 m isobaths for the northern and southern Wushi Fishery Port and for the Lanyang River estuary surveyed in 2006, 2013, 2013, and 2019 were collected and compared. Figure 3a,b illustrate the 0 m isobaths along the northern and southern Wushi Fishery Port, respectively, and the 0 m isobath for the Lanyang River estuary is depicted in Figure 3c. As shown in Figure 3a, the 0 m isobath extended offshore to the north of the Wushi Fishery Port by 2013 due to the construction of the jetty and groin. The 0 m isobath in the southern Wushi Fishery Port stretched offshore and reached its maximum in 2013, after which erosion occurred until 2015 (as shown in Figure 3b). The 0 m isobaths crossing the Lanyang River estuary in various years are shown in Figure 3c. The 0 m isobaths around the Lanyang River estuary trended offshore by 2012 because of abundant sediments supplied by the Lanyang River; however, they retreated landward until 2015 as a result of the river mouth moving northward and reduced sediment transport southward.

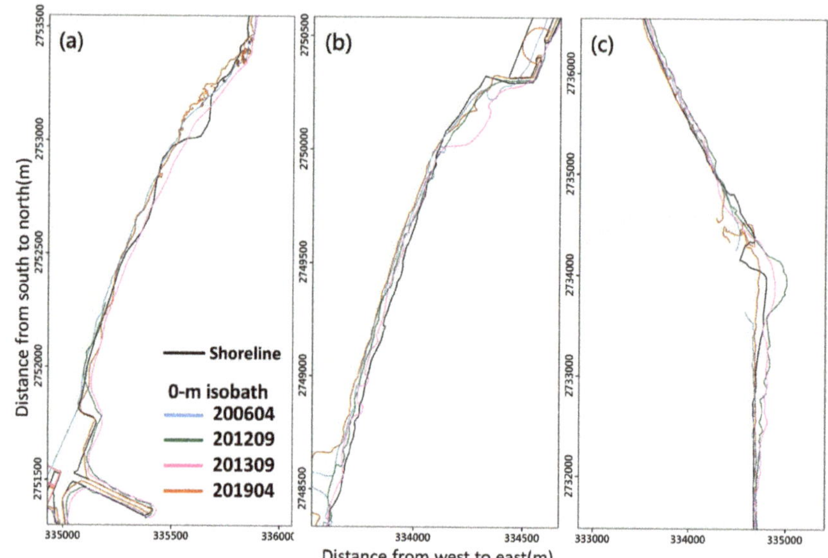

Figure 3. Evolution of the 0 m isobath along (**a**) the northern Wushi Fishery Port, (**b**) the southern Wushi Fishery Port, and (**c**) the Lanyang River estuary.

2.3. On-Site Bathymetric and Topographic Surveys

Multiple methods can be used for bathymetric surveys, e.g., multibeam and single-beam surveys. An accurate bathymetric survey allows the researcher to measure the depth of a water body and map the underwater features of a water body. To analyze the long-term erosion and accretion in the study area, topographic and bathymetric surveys were conducted through the measurement of 43 cross sections along the coastline from north of the Wushi Fishery Port to the south of the Lanyang River estuary in June 2006, April and September 2012, April and September 2013, and April and October 2019. Figure 4 demonstrates the spatial distribution of the cross sections for sampling bathymetry and topography. As shown in Figure 3, the planned survey track lines for the sonar collection were spaced approximately 500 m apart in the alongshore direction and were of varying length to allow the survey to be completed in 2 days. The bathymetric survey was performed offshore from north of the Wushi Fishery Port to the south of the Lanyang River estuary, utilizing a boat-mounted Global Positioning System (GPS) device, with a single-beam echosounder. The boat bathymetric survey was carried out until the measured water depth was approximately 10–15 m. This is based on the empirical formula proposed by Houston [36], i.e., *hc* = 6.75 *Hsm*, where *hc* is the closure depth (a water depth of the survey endpoint in nearshore waters) and *Hsm* is the long-term averaged significant wave height, which represents the wave climate of the study area. An *Hsm* value of 1.14 m was adopted in the present study, according to the statistical data for waves issued by the CWB of Taiwan. The *hc* value is estimated to be approximately 7.7 m for the nearshore waters of Yilan using Houston's equation. In order to avoid the inaccuracy of the depth sounder, the present study extended *hc* to 10–15 m to ensure the minimal *hc* (i.e., 7.7 m) is within the bathymetric surveys. A surveying-quality GPS unit and an electronic distance measurement (EDM) total station theodolite (TST) were used to determine the locations of features shown in the present study, to conduct a topographic survey on the beach.

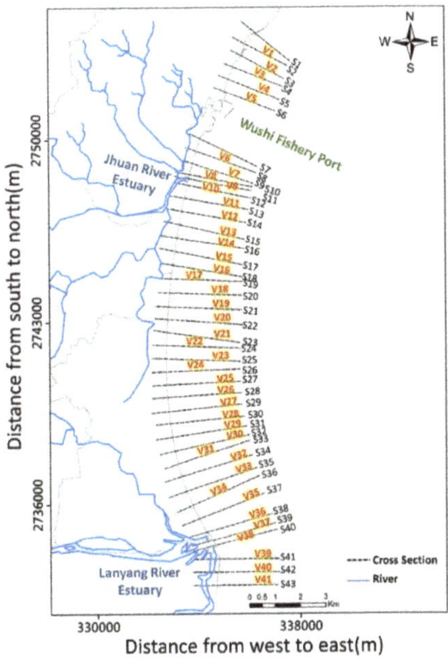

Figure 4. Spatial distribution of the cross sections for sampling bathymetry.

3. Results

3.1. Interannual Variability of Erosion and Accretion along the Surveyed Shoreline

To analyze the long-term variation in erosion and accretion for the study coastline, bathymetric surveys were conducted in June 2006, April and September 2012, April and September 2013, and April and October 2019. A total of 43 transects (track lines) were distributed along the shoreline from north of the Wushi Fishery Port to the south of the Lanyang River estuary (as shown in Figure 4). Moreover, the net volume of erosion or accretion between two transects was estimated, which means that 41 net volumes were used to evaluate whether the sediment budget was increasing or decreasing along the studied coast. The updated bathymetric datasets were compared with the previous datasets to estimate depth changes during two periods. Figure 5a,b present the variations in bottom elevation over 7 and 6 years, respectively, with Figure 5a showing the changes between June 2006 and April 2013 and Figure 5b showing the changes between April 2013 and April 2019.

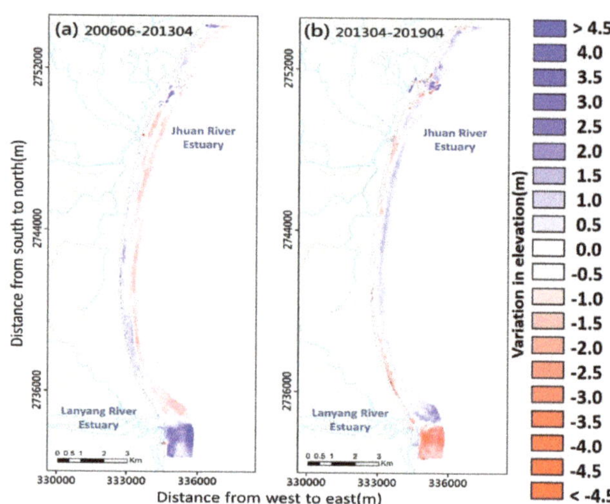

Figure 5. The spatial distribution of erosion and accretion along the shoreline during the periods of (**a**) June 2006 to April 2013 and (**b**) April 2013 to April 2019.

The comparisons indicate that erosion and accretion phenomena occurred alternately along the shoreline from north of the Wushi Fishery Port to the south of the Lanyang River estuary and from June 2006 to April 2013 and April 2013 to April 2019. This is particularly obvious in the waters near the Lanyang River estuary. As shown in Figure 5a, the bathymetries rose to a maximum of 3.0–4.0 m in the south of the Lanyang River estuary from June 2006 to April 2013; however, the water depths were reduced by 3.0–4.0 m in the same area from April 2013 to April 2019 (Figure 5b). Overall, the erosion and accretion trends for the waters close to the shoreline are found to be contrary to those of the water slightly farther from the shoreline for both periods. A similar phenomenon is discovered when Figure 5a is compared to Figure 5b. Slight erosion and accretion were distributed in the area somewhat far from the shoreline and the area close to the shoreline, respectively, along the coastline from north of the Wushi Fishery Port to the northern Lanyang River estuary during the period of June 2006 to April 2013 (i.e., a 7-year bottom elevation difference, as shown in Figure 5a). However, the opposite phenomenon of erosion and accretion occurred for the 6-year bottom elevation difference from April 2013 to April 2019 (Figure 5a). Figure 6a,b present the net erosion (positive quantity in Figure 6) or accretion (negative quantity in Figure 6) volume of each pair of transects in the 7-year and 6-year periods, respectively. Based on a comparison of Figure 5 with Figure 6, the distribution pattern of increases and decreases in the net volume is identical to that of the bathymetric changes alongshore. The maximal accretion volumes are approximately 2×10^6 m^3 at two intervals between transects S40 and S41 and between S41 and S42 (i.e., V39 and V40, as shown in Figure 6a), while the maximal erosion volume is nearly -2×10^6 m^3 at an interval between transects S41 and S42 (i.e., V40, as shown in Figure 6b). To evaluate the longer-term erosion and accretion along the studied shoreline, the differences between the bathymetric surveys in June 2006 and April 2019 were calculated. The variations in bottom elevation and net volume are depicted in Figure 6a,b, respectively. The 13-year alongshore erosion and accretion variations are minor relative to the 7-year and 6-year evolution. The changes in water depth are within ±2.0 m (as shown in Figure 7a), and the maximum accretion and erosion volumes are approximately 1.0×10^6 m^3 and -5.0×10^5 m^3 at V39 and V35, respectively (Figure 7b). This means that the sediment added to and removed from the studied coastal system is gradually balanced through long-term sediment transport.

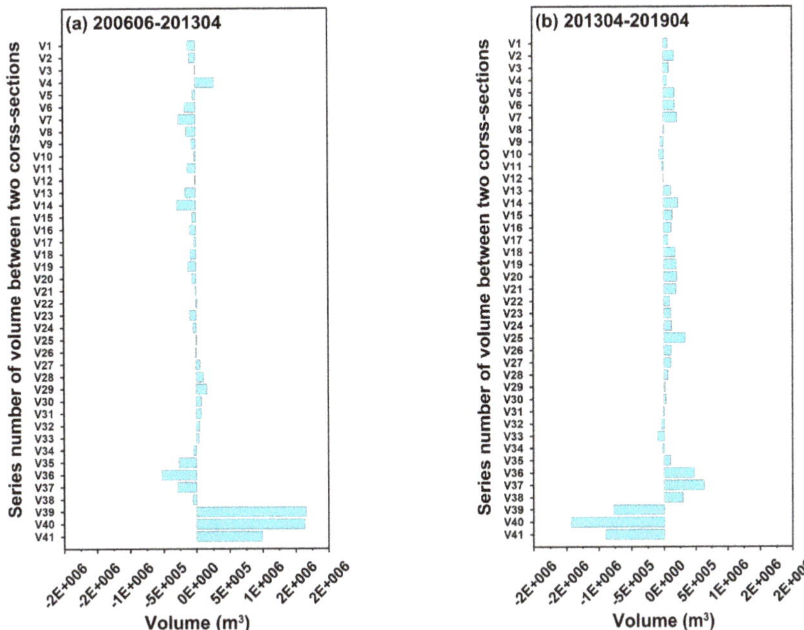

Figure 6. The erosion and accretion volume between each pair of cross sections during the period of (**a**) June 2006 to April 2013 and (**b**) April 2013 to April 2019.

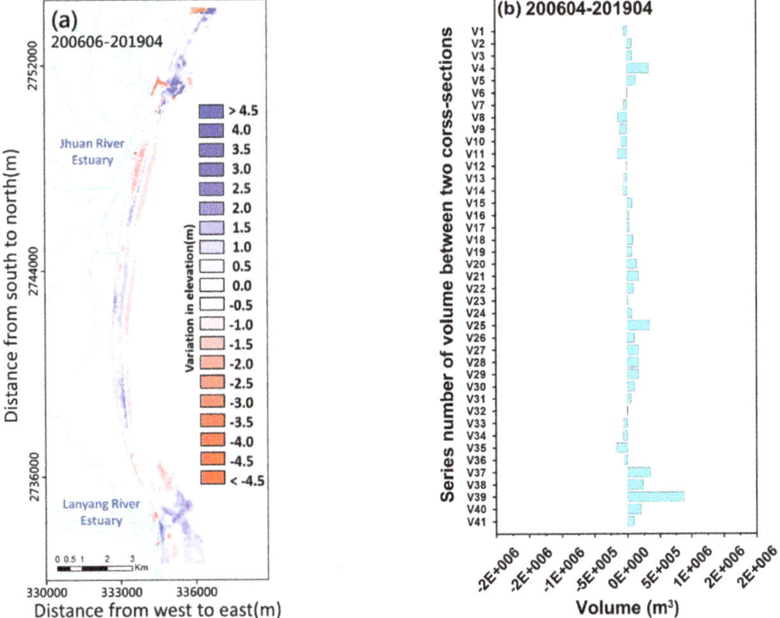

Figure 7. The erosion and accretion volume between each pair of cross sections during the period of (**a**) June 2006 to April 2013 and (**b**) April 2013 to April 2019.

3.2. Seasonal Variability in Erosion and Accretion along the Surveyed Shoreline

In addition to the interannual variability in erosion and accretion, estimating the seasonal variation in erosion and accretion is also important to coastal management and development. The bathymetric data pairs surveyed in September 2012 and April 2013 and in April 2013 and October 2013 were adopted to analyze the variability in erosion and accretion along the studied coastline in winter and summer, respectively. The bathymetry differences between September 2012 and April 2013 are regarded as erosion and accretion variations in winter (as shown in Figure 8a), while the bathymetry differences between April 2013 and October 2013 are considered to be erosion and accretion variations in summer (as shown in Figure 8b). As shown in Figure 8a, erosion phenomena are obvious along the shoreline from north of the Wushi Fishery Port to the south of the Lanyang River estuary during winter, due to the sustained large waves caused by the northeast monsoon and the shortage of sediment discharged from the river. In contrast, in summer, accretion phenomena are found along the coastline from north of the Wushi Fishery Port to the Lanyang River estuary (Figure 8b) because of weaker waves and abundant sediment released from the river.

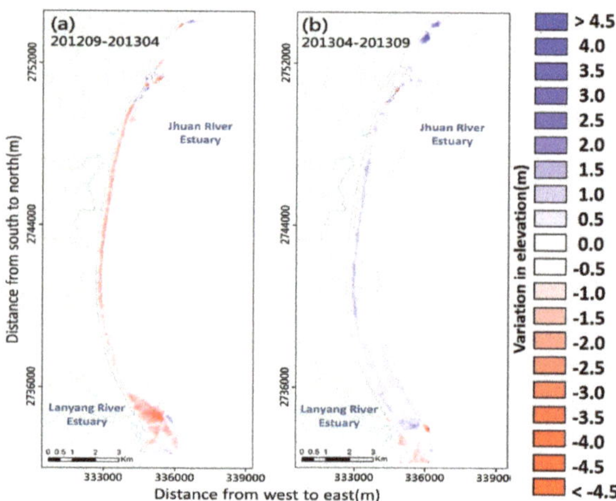

Figure 8. The spatial distribution of seasonal variability for erosion and accretion along the shoreline for (**a**) winter (from September 2012 to April 2013) and (**b**) summer (from April 2013 to September 2013).

3.3. Analysis of Cross-Shore Profile Evolution

Eight representative transects illustrated in Figure 9 were selected for cross-profile comparisons with the bathymetric surveys in June 2006, April 2012, April 2013, and April 2019. Cross-profile comparisons are beneficial for investigating nearshore morphological evolution and sediment transport mechanisms in coastal environments. Closure depth has been widely used within coastal engineering as an empirical measure of the seaward limit of significant cross-shore sediment transport on sandy beaches. Therefore, a closure depth of approximately 10 m was adopted as an offshore boundary for morphodynamics in the present study. The cross-profile surveys extended from dune ridges on the beach to the 10 m closure depth in the nearshore waters of the study site. Eight transects were measured in the northern and southern portions of the Wushi Fishery Port (transects S5 and S9 in Figure 9), in the northern and southern portions of the Jhuan River estuary (transects S10 and S11 in Figure 9), along the shoreline between the Jhuan River estuary and the Lanyang River estuary (transects S18 and S31 in Figure 9), and in the northern and southern portions of the Lanyang River estuary (transects S38 and S41 in Figure 9). Figure 10a–d represent the topographic and bathymetric data for transects S5 (Figure 10a), S9 (Figure 10b), S10

(Figure 10c), and S12 (Figure 10d). These transects were surveyed in June 2006, April 2012, April 2013, and April 2019. The cross-profile comparisons of transects S5 and S9 (surveyed in the northern and southern portions of the Wushi Fishery Port) in various years are shown in Figure 10a,c. The bathymetric surveys were all performed after the construction of the Wushi Fishery Port in 2002. The comparisons show that the accretion is sustainable in the northern portion of the Wushi Fishery Port (transect S5) and is up to 2 m at a distance of approximately 400 m from 2006 to 2019 (Figure 10a). Erosion phenomena were found in the southern portion of the Wushi Fishery Port (transect S9) at a distance of 100–400 m; however, weak accretion was detected at a distance of 600–800 m (Figure 10b). The construction of the Wushi Fishery Port, which created a jetty effect, is the major contributing factor to accretion and erosion on the northern and southern sides of the Wushi Fishery Port. The bathymetry changes are minor below a water depth below 0 m on the northern and southern sides of the Jhuan River estuary (transect S10 in Figure 10c and transect S12 in Figure 10d). However, a coastal dune with a height of approximately 7 m in transect S10 (northern side of the Jhuan River estuary) was eroded significantly within a distance of 150 m because of the strong northeast monsoon (Figure 10c). Transects S18 and S31 lay between the Jhuan River estuary and the Lanyang River estuary and feature a gently sloping seafloor (as shown in Figure 11a,b). The cross-profile comparisons are relatively stable because they are far from the estuaries and any artificial structures. Figure 11c,d illustrate the bathymetric variations along transects S38 and S41, respectively, in various years. As shown in Figure 11c (transect S38), erosion occurred in the north of the Lanyang River estuary with a distance between 700 m and 1300 m from 2006 to 2013, after which accretion was present in the same zone. The most significant erosion and accretion are observed in the southern Lanyang River estuary, i.e., transect S41 in Figure 11d. The accretion reached 2–4 m along all of transect S41 from 2006 to 2013, but the coastal and nearshore zones at distances beyond 200 m were eroded dramatically in 2019. These phenomena are identical to the evolution of the 0 m isobath in various years described in Section 2.2. The Lanyang River mouth moved northward gradually, and the supply of riverine sediment for the southern Lanyang River estuary has consequently decreased in the past two decades.

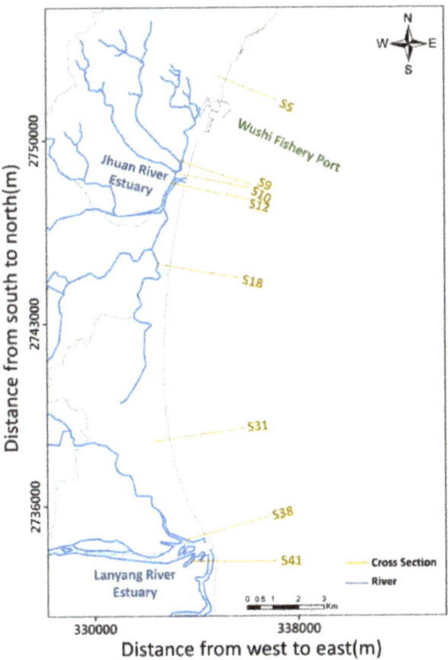

Figure 9. Spatial distribution of eight transects for the cross-shore profile comparison.

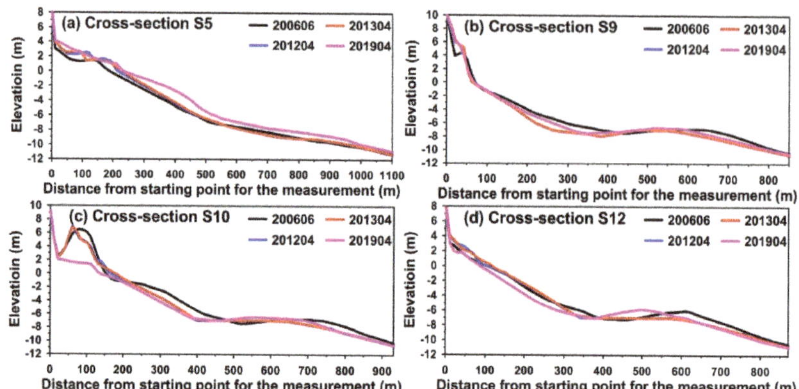

Figure 10. Cross-shore bathymetry profile comparisons for (**a**) S5, (**b**) S9, (**c**) S10, and (**d**) S12 cross-sections measured in various years.

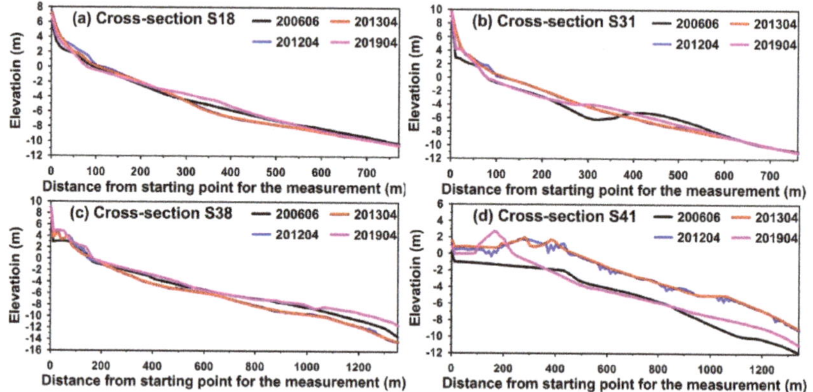

Figure 11. Cross-shore bathymetry profile comparisons for (**a**) S18, (**b**) S31, (**c**) S38, and (**d**) S41 cross sections measured in various years.

4. Discussion

According to the evolution of the 0 m isobath and the cross-sectional bathymetric data surveyed in various years, the present study reveals that a drastic morphodynamic pattern is found around the Lanyang River estuary and is particularly obvious in the southern Lanyang River estuary. The topographic and bathymetric data of transects S30–S43 surveyed in June 2006, April 2013, and April 2019 were utilized to clarify the mechanism of estuarine sediment transport for the Lanyang River. The spatial distributions of transects S30–S43 and their corresponding serial numbers for the net volume variation (i.e., V29–S43) are delineated in Figure 12a for June 2006 to April 2013 and in Figure 12b for April 2013 to April 2019. A comparison of Figure 12a,b shows that the erosion and accretion phenomena in the northern Lanyang River estuary are contrary to those in the southern Lanyang River estuary for the two periods. This is because a nature reserve in the southern Lanyang River estuary cannot be developed from 2013. Although the 13-year (June 2006 to April 2019) variations still show a slight accretion in the south of the Lanyang River estuary (Figure 7), erosion began in 2013. The erosion and accretion along the shoreline near the Lanyang River estuary are highly dependent on the estuarine sediment discharge and transport of the Lanyang River. Therefore, the limit of estuarine sediment transport is considered to play an important role in the mechanism of coastal erosion and accretion and can be used to examine how far estuarine sediment can be supplied. The net volume variations along the coastline of the Lanyang River estuary for the periods from June 2006 to April 2013 (cyan bar) and from April 2013 to April 2019 (red bar) are presented in Figure 12. As shown in Figure 13, erosion and accretion began to reverse at V34 and V38 between the two periods. The accretion turned into erosion from V34 to V38 and then returned to accretion at V39 during the 7-year period (from June 2006 to April 2013, cyan bar in Figure 13). A contrary phenomenon appeared in the same interval (between V34 and V38) in the 6-year period (from April 2013 to April 2019, red bar in Figure 13); in other words, accretion occurred within V34 to V38, but erosion occurred outside this interval. Therefore, it is believed that the extent affected by estuarine sediment is approximately 2 km long from north to south along the coast of the Lanyang River estuary.

Infrastructure and human activity, such as establishing nature reserves and constructing jetties along the coast, modify many of the shorelines worldwide and drive change in coastal geomorphology. These artificial modifications resulted in impairment to the balance of the erosion and accretion in the estuarine and coastal environment. The on-site data surveyed in the present study provides insightful and valuable suggestions in scientific perspectives and coastal zone management.

The extensive set of observed data in the present study are undoubtedly helpful for future comparisons with the results from numerical models to determine the beach shoreline evolution [37,38]; in fact, monitoring activities significantly contribute to improving the robustness and reliability of numerical models [39,40].

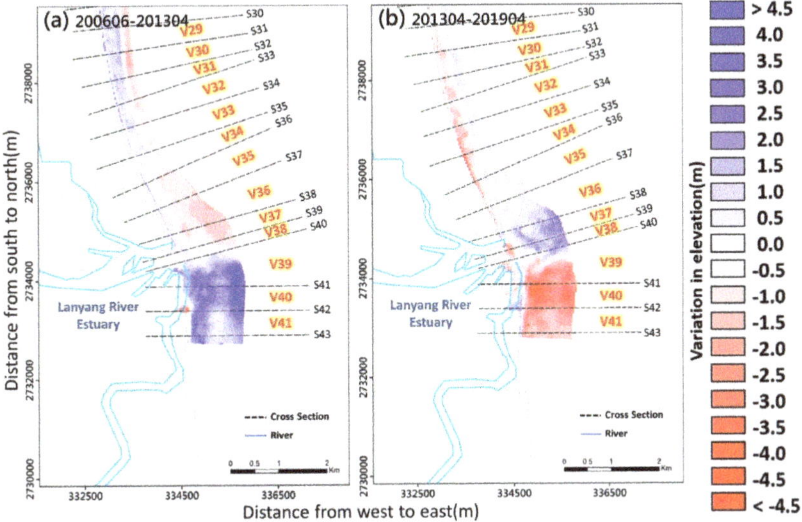

Figure 12. Long-term erosion and accretion and spatial distribution of the cross sections for sampling bathymetry near the Lanyang River estuary during the period of (**a**) June 2006 to April 2013 and (**b**) April 2013 to April 2019.

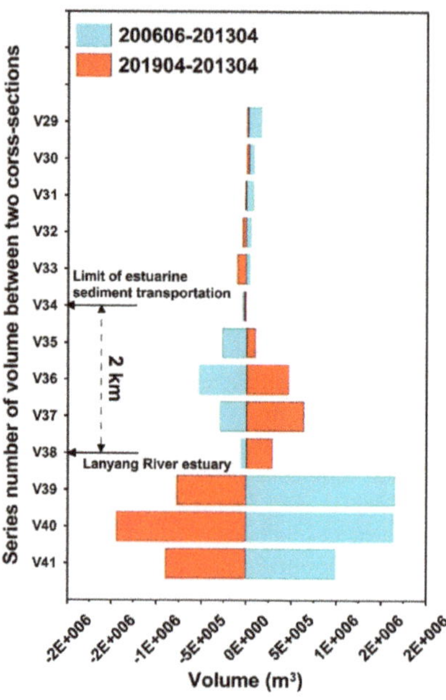

Figure 13. Net volume variations between erosion and accretion around the Lanyang River estuary (corresponding to V29 to V41 in Figure 11) during the period of June 2006 to April 2013 (cyan bar) and April 2013 to April 2019 (red bar).

5. Summary and Conclusions

A series of on-site topographic and bathymetric surveys were conducted along a 20 km long stretch of shoreline from north of the Wushi Fishery Port to the south of the Lanyang River estuary in June 2006, April and September 2012, April and September 2013, and April and October 2019. A total of 43 planned survey track lines designed for sonar collection were spaced approximately 500 m apart in the alongshore direction. Each track line was of varying length to allow the maximum water depth to reach 10–15 m (deeper than the closure depth), and topographic and bathymetric surveys were completed in 2 days. The bathymetric data obtained from several on-site surveys and the collected 0 m isobaths in various years were employed to investigate the long-term erosion and accretion along the studied shoreline. The 0 m isobath evolution reveals that the construction of a jetty and groin for the Wushi Fishery Port caused erosion and accretion to occur to the north and south, respectively. The interannual comparisons of the alongshore and cross-shore profiles indicate that dramatic erosion occurred in the southern Lanyang River estuary from 2013 to 2019 due to the establishment of a nature reserve. However, slight accretion occurred from 2013 to 2019 in the northern Lanyang River estuary because the riverine sediments were carried northward by tidal currents. In winter, the strong northeast monsoon induced sustained large waves, and the shortage of riverine sediment supply led to erosion of the coastal areas of northeastern Taiwan. In contrast to winter, weaker waves and large amounts of sediment released from the river (because of high river discharge) in summer were beneficial for accretion in the coastal areas of northeastern Taiwan. The cross-profile evolutions indicate that the supply of riverine sediment in the southern Lanyang River estuary was reduced in the past two decades as a result of the gradual northward movement of the Lanyang River mouth. The shoreline affected by estuarine sediment is approximately 2 km long from north to south of the Lanyang River alongshore. Long-term alongshore erosion and accretion are natural processes in coastal areas; however, the loss of coastal lands could be a result of overexploitation, e.g., the designs and constructions of the fishery port and backwater for the coastal areas. Although the present study provides insight into the problems of management and development for the coastal environment, more frequent topographic and bathymetric surveys for the whole shoreline of Taiwan should be conducted in the future. Additionally, numerical simulations are necessary for future research to explain the seasonal mechanism of the simulations of the seasonal and annual change for wavefield, tidal current, and alongshore sediment transport.

Author Contributions: Conceptualization, T.-Y.L., S.-C.H., C.-H.C., T.-Y.C., W.-B.C., W.-P.H., W.-D.G., J.-Y.H. and C.-H.L.; on-site surveys, T.-Y.L., C.-H.C., T.-Y.C., W.-B.C., W.-P.H., W.-D.G., J.-Y.H. and C.-H.L.; data collection and analysis, T.-Y.L., C.-H.C., T.-Y.C., W.-P.H. and W.-B.C.; writing—original draft preparation, W.-B.C. and S.-C.H.; writing—review and editing, W.-B.C. and S.-C.H. All authors have read and agreed to the published version of the manuscript.

Funding: This research was supported by the National Science and Technology Center for Disaster (NCDR), Ministry of Science and Technology (MOST), Taiwan.

Institutional Review Board Statement: Not applicable.

Informed Consent Statement: Not applicable.

Data Availability Statement: Publicly available datasets were analyzed in the present study. The data can be found here: https://ocean.cwb.gov.tw/V2/ (accessed on 5 January 2022).

Acknowledgments: The authors would like to thank the Central Weather Bureau and the Water Resource Agency, Taiwan, for providing the measurements of waves and currents.

Conflicts of Interest: The authors declare no conflict of interest.

References

1. Rueda, A.; Vitousek, S.; Camus, P.; Tomás, A.; Espejo, A.; Losada, I.J.; Barnard, P.L.; Erikson, L.H.; Ruggiero, P.; Reguero, B.G.; et al. A global classification of coastal flood hazard climates associated with large-scale oceanographic forcing. *Sci. Rep.* **2017**, *7*, 5038. [CrossRef] [PubMed]
2. Mentaschi, L.; Vousdoukas, M.I.; Pekel, J.-F.; Voukouvalas, E.; Feyen, L. Global long-term observations of coastal erosion and accretion. *Sci. Rep.* **2018**, *8*, 12876. [CrossRef] [PubMed]
3. Pang, J.Z.; Si, S.H. Evolution of the Yellow River mouth: I. Historical shifts. *Oceanol. Limnol. Sin.* **1979**, *10*, 136–141.
4. Pang, J.Z.; Jiang, M.X.; Li, F.L. Changes and development trend of runoff, sediment discharge and coastline of the Yellow River Estuary. *Trans. Oceanol. Limnol.* **2000**, *4*, 3–6.
5. Wright, L.D.; Wiseman, W.J.; Bornhold, B.D.; Prior, D.B.; Suhayda, J.N.; Keller, G.H.; Yang, Z.S.; Fan, Y.B. Marine dispersal and deposition of Yellow River silts by gravity-driven underflows. *Nature* **1988**, *332*, 629–632. [CrossRef]
6. Blodget, H.W.; Taylor, P.T.; Roark, J.H. Shoreline changes along the Rosetta-Nile Promontory monitoring with satellite observations. *Mar. Geol.* **1991**, *99*, 67–77. [CrossRef]
7. Frihy, O.E.; Komar, P.D. Long-term shoreline changes and the concentration of heavy minerals in beach sands of the Nile Delta, Egypt. *Mar. Geol.* **1993**, *115*, 253–261. [CrossRef]
8. Huang, H.J.; Li, C.Z.; Guo, J.J. Application of Landsat images to the studies of the shoreline changes of the Huanghe River Delta. *Mar. Geol. Quat. Geol.* **1994**, *14*, 29–37.
9. Huang, H.J.; Wang, Z.Y.; Zhang, R.S. The error analysis of the methods of monitoring the beach changes using digital photogrammetry, satellite images and GIS. *Mar. Sci.* **2002**, *26*, 8–10.
10. Ji, Z.W.; Hu, C.H.; Zeng, Q.H. Analysis of recent evolution of the Yellow River Estuary by Landsat images. *J. Sediment. Res.* **1994**, *3*, 12–22.
11. Xue, C.T. Historical changes in the Yellow River Delta, China. *Mar. Geol.* **1993**, *113*, 321–329. [CrossRef]
12. Xue, C.T. Division and recognition of modern Yellow River Delta lobes. *Geogr. Res.* **1994**, *13*, 59–66.
13. Sun, X.G.; Yang, Z.S. Prediction of modern Huanghe River Delta area increment by using the sediment discharge. *Oceannol. Limnol. Sin.* **1995**, *26*, 76–82.
14. Li, G.X.; Wei, H.L.; Han, Y.S.; Cheng, Y.J. Sedimentation in the Yellow River Delta: Part I. Flow and suspended sediment structure in the upper distributary and the estuary. *Mar. Geol.* **1998**, *149*, 93–111. [CrossRef]
15. Li, F.L.; Pang, J.Z.; Jiang, M.X. Shoreline changes of the Yellow River Delta and its environmental geology effect. *Mar. Geol. Quat. Geol.* **2000**, *20*, 17–21. (In Chinese)
16. White, K.; Asmar, H.M. Monitoring changing position of coastlines using Thematic Mapper imagery, an example from the Nile Delta. *Geomorphology* **1999**, *29*, 93–105. [CrossRef]
17. Dias, J.M.A.; Boski, T.; Rodrigues, A.; Magalhaes, F. Coastline evolution in Portugal since the Last Glacial Maximum until present—A synthesis. *Mar. Geol.* **2000**, *170*, 177–186. [CrossRef]

18. Xu, J.X. A study of thresholds of runoff and sediment for the land accretion of the Yellow River Delta. *Geogr. Res.* **2002**, *21*, 163–170.

19. Yu, L.S. The Huanghe (Yellow) River: A review of its development, characteristics, and future management issues. *Cont. Shelf Res.* **2002**, *22*, 389–403. [CrossRef]

20. Chang, J.; Liu, G.H.; Liu, Q.S. Analysis on spatio-temporal feature of coastline change in the Yellow River Estuary and its relation with runoff and sand-transportation. *Geogr. Res.* **2004**, *23*, 339–346.

21. Syvitski, J.P.M.; Vorosmarty, C.J.; Kettner, A.J.; Green, P. Impact of humans on the flux of terrestrial sediment to the global coastal ocean. *Science* **2005**, *308*, 376–380. [CrossRef] [PubMed]

22. Murray, N.J.; Phinn, S.R.; DeWitt, M.; Ferrari, R.; Johnston, R.; Lyons, M.B.; Clinton, N.; Thau, D.; Fuller, R.A. The global distribution and trajectory of tidal flats. *Nature* **2019**, *565*, 222–225. [CrossRef] [PubMed]

23. Chen, W.-B.; Liu, W.-C.; Hsu, M.-H.; Hwang, C.-C. Modelling investigation of suspended sediment transport in a tidal estuary using a three-dimensional model. *Appl. Math. Model.* **2015**, *39*, 2570–2586. [CrossRef]

24. Nienhuis, J.H.; Ashton, A.D.; Edmonds, D.A.; Hoitink, A.J.F.; Kettner, A.J.; Rowland, J.C.; Törnqvist, T.E. Global-scale human impact on delta morphology has led to net land area gain. *Nature* **2020**, *577*, 514–518. [CrossRef]

25. Dunn, F.; Darby, S.; Nicholls, R.; Cohen, S.; Zarfl, C.; Fekete, B. Projections of declining fluvial sediment delivery to major deltas worldwide in response to climate change and anthropogenic stress. *Environ. Res. Lett.* **2019**, *14*, 1–11. [CrossRef]

26. Warrick, J.A.; Stevens, A.W.; Miller, I.M.; Harrison, S.R.; Ritchie, A.C.; Gelfenbaum, G. World's largest dam removal reverses coastal erosion. *Sci. Rep.* **2019**, *9*, 13968. [CrossRef] [PubMed]

27. Hapke, C.J.; Kratzmann, M.G.; Himmelstoss, E.A. Geomorphic and human influence on large-scale coastal change. *Geomorphology* **2013**, *199*, 160–170. [CrossRef]

28. Angamuthu, B.; Darby, S.E.; Nicholls, R.J. Impacts of natural and human drivers on the multi-decadal morphological evolution of tidally-influenced deltas. *Proc. R. Soc. A* **2018**, *474*, 20180396. [CrossRef] [PubMed]

29. Hurst, M.D.; Rood, D.H.; Ellis, M.A.; Anderson, R.S.; Dornbusch, U. Recent acceleration in coastal cliff retreat rates on the south coast of Great Britain. *Proc. Natl. Acad. Sci. USA* **2016**, *113*, 13336–13341. [CrossRef] [PubMed]

30. Dadson, S.; Hovius, N.; Chen, H.; Dade, W.B.; Hsieh, M.-L.; Willett, S.D.; Hu, J.-C.; Horng, M.-J.; Chen, M.-C.; Stark, C.P.; et al. Links between erosion, runoff variability and seismicity in the Taiwan orogen. *Nature* **2003**, *426*, 648–651. [CrossRef] [PubMed]

31. United Nations International Strategy for Disaster Reduction (UNISDR). In *Global Survey of Early Warning Systems: An Assessment of Capacities, Gaps and Opportunities towards Building a Comprehensive Global Early Warning System for All Natural Hazards*; United Nations International Strategy for Disaster Reduction: Geneva, Switzerland, 2006.

32. Leaman, C.K.; Harley, M.D.; Splinter, K.D.; Thran, M.C.; Kinsela, M.A.; Turner, I.L. A storm hazard matrix combining coastal flooding and beach erosion. *Coast. Eng.* **2021**, *170*, 104001. [CrossRef]

33. Romagnoli, C.; Sistilli, F.; Cantelli, L.; Aguzzi, M.; De Nigris, N.; Morelli, M.; Gaeta, M.G.; Archetti, R. Beach Monitoring and Morphological Response in the Presence of Coastal Defense Strategies at Riccione (Italy). *J. Mar. Sci. Eng.* **2021**, *9*, 851. [CrossRef]

34. Shih, H.-J.; Chen, H.; Liang, T.-Y.; Fu, H.-S.; Chang, C.-H.; Chen, W.-B.; Su, W.-R.; Lin, L.-Y. Generating potential risk maps for typhoon-induced waves along the coast of Taiwan. *Ocean Eng.* **2018**, *163*, 1–14. [CrossRef]

35. Chang, C.-H.; Shih, H.-J.; Chen, W.-B.; Su, W.-R.; Lin, L.-Y.; Yu, Y.-C.; Jang, J.-H. Hazard Assessment of Typhoon-Driven Storm Waves in the Nearshore Waters of Taiwan. *Water* **2018**, *10*, 926. [CrossRef]

36. Houston, J.R. Simplified Dean's method for beach-fill design. *J. Waterw. Port Coast. Div.* **1996**, *122*, 143–146. [CrossRef]

37. Hanson, H.; Kraus, N.C. Long-Term Evolution of a Long-Term Evolution Model. *J. Coast. Res.* **2011**, *SI 59*, 118–129. [CrossRef]

38. Frey, A.E.; Connell, K.J.; Hanson, H.; Larson, M.; Thomas, R.C.; Munger, S.; Zundel, A. *GenCade Version 1 Model Theory and User's Guide*; Technical Report ERDC/CHL TR-12-25; U.S. Army Engineer Research and Development Center: Vicksburg, MS, USA, 2012.

39. Medellín, G.; Torres-Freyermuth, A.; Tomasicchio, G.R.; Francone, A.; Tereszkiewicz, P.A.; Lusito, L.; Palemón-Arcos, L.; López, J. Field and Numerical Study of Resistance and Resilience on a Sea Breeze Dominated Beach in Yucatan (Mexico). *Water* **2018**, *10*, 1806. [CrossRef]

40. Puleo, J.A.; Cristaudo, D.; Torres-Freyermuth, A.; Masselink, G.; Shi, F. The role of alongshore flows on inner surf and swash zone hydrodynamics on a dissipative beach. *Cont. Shelf Res.* **2020**, *201*, 104134. [CrossRef]

MDPI

St. Alban-Anlage 66

4052 Basel

Switzerland

Tel. +41 61 683 77 34

Fax +41 61 302 89 18

www.mdpi.com

Journal of Marine Science and Engineering Editorial Office

E-mail: jmse@mdpi.com

www.mdpi.com/journal/jmse